Social To'

ONE WEEK LOAN

Many issues – such as access for the disa
matters and ethnic minority issues – are ex
tions by planning authorities. This book in
planning' in order to integrate planning poli
social issues of the people they are planning f

The first part of the book provides backgro
dimension to the predominantly physical, land
system. There follows an investigation of a rep
planning topics in respect of gender, race, age a
implications for mainstream policy areas such a

D0493757

transport. The next part discusses the likely influence of a range of global and
European policy initiatives and organisations in changing the agenda of British
town planning. Planning for healthy cities, sustainability, social cohesion and
equity are discussed. The book concludes with a look at 'the problem' from a cultural perspective, arguing that a great weakness in the British system, resulting in
ugly and impractical urban design, has been the lack of concern among planners
with social activities and cultural diversity. Alternative, more culturally inclusive
approaches to planning, such as urban time planning, are presented which might
transcend the social/spatial dichotomy.

Concluding that the process of planning must change, the authors argue that
the culture and composition of the planning profession itself must change, particularly to be more representative and reflective of the people they are 'planning
for', in terms of gender, race and minority composition.

Clara H. Greed is a Reader in the Faculty of the Built Environment, University
of the West of England, Bristol.

Social Town Planning

Edited by Clara H. Greed

London and New York

First published 1999
by Routledge
2 Park Square, Milton Park, Abingdon, Oxon, OX14 4RN

Simultaneously published in the USA and Canada
by Routledge
270 Madison Ave, New York NY 10016

Transferred to Digital Printing 2006

Typeset in Galliard by
M Rules

British Library Cataloguing in Publication Data
A catalogue record for this book is available from the British Library

Library of Congress Cataloging in Publication Data
Social town planning: planning and social policy/edited by
  Clara H. Greed.
    p.   cm.
  Includes bibliographical references and index.
  1. City planning – Social aspects – Great Britain.   2. City planning
  – Social aspects – Europe.   3. Minorities – Services for – Great
  Britain.   4. Minorities – Services for – Europe.   5. Architecture
  and the handicapped – Great Britain.   6. Architecture and the
  handicapped – Europe.   I. Greed, Clara.
  HT169.G7S64   1999
  307.1'.216'0941–dc21   98-43905

ISBN 0–415–17240–3 (hbk)
ISBN 0–415–17241–1 (pbk)

Publisher's Note
The publisher has gone to great lengths to ensure the quality of this reprint
but points out that some imperfections in the original may be apparent

Printed and bound by CPI Antony Rowe, Eastbourne

Contents

Illustrations

Figures

Tables

Boxes

Contributors

Professor Franco Bianchini is Reader in Cultural Planning and Policy and programme leader for the MA in European Cultural Planning at de Montfort University, Leicester. Drawing on Italian initiatives, he has been involved in the development of the concept of 'time planning' in Britain. His publications include *Out of Hours: A Study of Economic, Social and Cultural Life in Twelve Town Centres in the UK* (with colleagues from the Comedia Group, London, 1991), and *The Creative City* (with C. Landry, published by Demos, London). He has worked as a researcher or advisor to the European Commission, the Council of Europe and various local authorities and academic institutions in Britain and Italy.

Janet Brand is a senior lecturer in the Department of Environmental Planning at the University of Strathclyde, Glasgow, Scotland. Her primary research and teaching interests relate to planning practice, particularly development plans and development control, and to sustainable development. In the 1980s she completed research for the Scottish Office on planning applications and planning appeals. Her research work, *Review of the Use Classes Order*, was published by the Scottish Office (1994). She has published widely on a range of planning issues and is listed in the Scottish *Who's Who* because of her high profile in planning and environmental activities.

Rob Davidson is a senior lecturer in tourism. He has recently joined the staff of the University of Westminster Department of Tourism after spending nine years in France teaching tourism at two universities and the ESSEC–Cornell Institute of Management in Paris. Prior to that he spent five years as Education and Training Manager with the British Tourist Authority in London. His main areas of expertise are European-wide tourism trends and business tourism. He has produced a range of publications on these themes. He holds an MA in English Literature and Language from the University of Aberdeen.

Linda Davies is a chartered town planner who combines professional and teaching activities and, in the process, seeks to increase equal opportunities awareness among planners. She is a senior lecturer at the University of the West of England, and also teaches at Reading University. Previously she spent over fifteen years working in inner urban Scottish local authorities, and in rural Wiltshire. She is an executive committee member and ethnic minority liaison officer for the south-west branch of the Royal Town Planning Institute, and maintains an interest in disability issues.

Colin Fudge is Dean of the Faculty of the Built Environment, University of the West of England, Bristol and was previously Director of Environmental Services at Bath City Council. He is Urban Environment Advisor to the European Commission (DG XI); advisor on research (DG XII); Chair of the EU Urban Environment Expert Group; Chair of the OECD Urban Indicators International Panel, and Director of the World Health Organisation Collaborative Research Centre on Healthy Cities and Urban Policy. He has published widely and is also involved in a range of research projects and initiatives internationally.

Rose Gilroy is a senior lecturer at the University of Newcastle. Her career has spanned the fields of drama, the arts and a range of cultural initiatives. She is a member of the steering group for Eurofem, the Europe-wide 'women and planning' organisation, and recently took a leading organisational role in the 1998 Eurofem conference in Finland. Her research interests have focused on exploring how the practices and policies of planning and housing can create and sustain oppression. She is a fellow of the Chartered Institute of Housing and her publications include *Housing Women* (Routledge, 1994, co-edited with Roberta Woods).

Ann de Graft-Johnson is an architect and a senior lecturer at the University of the West of England. She was a member of Matrix Architects Feminist Co-operative in London until its closure in 1996. She has worked on housing, special needs, community, educational, arts, childcare and urban regeneration projects and published widely. Ann is a founder member of the Society of Black Architects (SOBA). She was also a member of the working party of Black and Minority Ethnic construction professionals headed by Bernie Grant MP which produced the discussion document *Building E=quality* (1996).

Clara H. Greed is a reader in the Faculty of the Built Environment, University of the West of England, Bristol. Her current research is centred on developing themes from ESRC research undertaken in 1997, on the changing culture and composition of the construction industry, which has widened out from her previous work on the surveying and town planning professions. She has published widely, producing research monographs, such as *Women in Planning: Creating Gendered Realities* (Routledge, 1994) and a series of textbooks, including *Introducing Town Planning* (Longmans, 1996) aimed at changing the hearts and minds of planning students.

Jean Hillier left the UK for Australia at the end of 1988, having previously worked both in planning practice and education, contributing to the establishment of the 'women and planning' movement in the 1980s. She is Associate

Professor in the Department of Urban and Regional Planning at Curtin University in Perth, Western Australia. She is a core group member of the Western Australia, Women in Planning Network. She has researched and published widely on citizen participation in planning, social planning, planning theory and on ethics and values in planning praxis.

Jo Little is a lecturer in geography at the University of Exeter. Her interests include rural issues, gender, and the implementation of planning policy. Her publications include *Gender, Planning and the Policy Process* (Elsevier Pergamon, 1994), which highlighted the problem of an abundance of policy statements but limited evidence of implementation in respect of 'planning for women'. She co-edited (with Paul Cloke) *Contested Countryside Cultures: Otherness, Marginalisation and Rurality* (Routledge, 1997). She is a member of the Women and Geography Study Group of the Royal Geographical Society/Institute of British Geographers. More recently she has undertaken ESRC-funded research on the Rural Challenge Initiative.

Robert Maitland is Chair of the Department of Tourism at the University of Westminster, London. He holds degrees in economics, environmental planning and business administration. He has worked on economic policy development in local government, and was head of a research management consultancy. His research interests are tourism destination management, on which he has advised the British government, and urban regeneration initiatives. He is undertaking a research project on the effectiveness of UK government policy for leisure and tourism for the DETR. He has published with Rob Davidson *Tourism Destinations* (Hodder & Stoughton, 1997).

Nigel Taylor is a principal lecturer in the Faculty of the Built Environment, University of the West of England, Bristol. He describes himself primarily as a philosopher and has pursued interests spanning planning theory and history, and urban design, as well as contributing to the development of a more reflective approach to planning education, and always making time to help individual students. He has published widely and has contributed, through his editorial roles, to the development of planning journal development. His most recent major publication is *Urban Planning Theory Since 1945* (Paul Chapman, 1998).

Huw Thomas is a professor at the Department of City and Regional Planning, University of Wales, Cardiff, having previously worked in local government and at Oxford Brookes University. His recent research, funded by ESRC, includes work on urban renewal, race equality and local governance. He has written on a range of social town planning issues, with particular emphasis upon race, and contributed to professional committees and community groups in this field. Publications include *Race, Equality and Planning: Policies and Procedures* (jointly edited with V. Krishnarayan, Avebury, 1994).

Geoff Vigar is a lecturer in the School of the Built Environment at Leeds Metropolitan University, having previously worked at the University of

Newcastle. His main research interests concern the relationship between environmentalism and governance; institutional approaches to policy analysis; and land-use planning and transport policy. He is co-author of *Planning, Place and Governance: Spatial Strategy-Making and the English Planning System* (Macmillan, forthcoming). He is principal investigator on a research project looking at the micro-politics of urban transport planning, funded by the Rees Jeffreys Road Fund.

Richard Williams is a member of the Royal Town Planning Institute and a senior lecturer at Newcastle University. He is a specialist in supra-national spatial policy and is an expert adviser to the Council of Europe and the OECD, a delegate to DG XVI seminars on ESDP and a member of the Scientific Council (Wissenschaftlichen Beirat) of the German Government's Federal Office for Building and Regional Planning (Bundesamt für Bauwesen und Raumordnung). He has held visiting professorships in Helsinki and Aalborg. He is author of *European Union Spatial Policy and Planning* (Paul Chapman, 1996) and many other related publications.

Acknowledgements

Chapter 6: thanks are given to Peter Monk, Access Officer for Thamesdown Borough Council, Swindon, for information on the role of the Building Regulations in access provision. Table 6.1 is, in part, based upon his advice, as subsequently developed by C. Greed and M. Roberts for use (in a non-tabular format) in *Introducing Urban Design* (Longmans, 1998, p. 234). The contribution of other access officers and disability groups is also gratefully acknowledged.

Figures 8.2 and 8.3 are from the Labo Housing Association scheme with copyright acknowledgment to the Home Housing Trust.

Chapter 10, by Colin Fudge, draws on the work of the EU Urban Environment Expert Group, its two policy reports, work for WHO on Healthy Cities and work for the European Commission DGs XI (Environment), XII (Research), and XVI (Regional). It acknowledges the many who contributed to this work, but particularly Agis Tsouros for permission to use material from *City Planning for Health and Sustainable Development* (WHO, 1997) in the city case studies, David Black, Liz Mills, Stefaan de Rynck, David Ludlow, Susann Pauli, Anthony Payne and Janet Rowe.

Janet Brand acknowledges the assistance of the following persons in the research undertaken to produce chapter 11: Alison Beyer, Chief Executive's Department, City of Glasgow Council; Kenny Boag, Environmental Health Department, South Lanarkshire Council and Convener of the Scottish Local Agenda 21 Co-ordinators' Network; George Chalmers, Executive Director, Forward Scotland; Kevin Dunion, Director, Friends of the Earth, Scotland; Phil Matthews, Secretary, Lord Provost's Commission on Sustainable Development, City of Edinburgh; Alistair McKenzie, Chief Planner, The Scottish Office Development Department; Janice Pauwels, Strategic Policy Department, City of Edinburgh Council; Geoff Pearson, Head of Sustainable Development Team, The Scottish Office; Graham U'Ren, Director of Policy, RTPI in Scotland.

Chapter 15: research undertaken on the construction industry entitled 'Social Integration and Exclusion in Professional Subcultures in Construction', referred to in this chapter, was undertaken in 1997 by C. Greed and funded by the ESRC, Project Number ROOO 22 1916.

Abbreviations

ABE	Association of Building Engineers
ACE	Access Committee for England
AGSD	Advisory Group on Sustainable Development
ARB	Architects' Registration Board
ARCOM	Association of Researchers in Construction Management
ARU	Les Annales de la Recherche Urbaine
ASI	Architects' and Surveyors' Institute
BCODP	British Council of Organisations of Disabled People
BS	Building Standards
BSI	British Standards Institution
CADISPA	Conservation and Development in Sparsely Populated Areas
CAE	Centre for Accessible Environments
CAP	Common Agricultural Policy
CBI	Confederation of British Industry
CCP	Capital City Partnership
CCT	closed-circuit television
CEC	Commission of the European Communities
CEMR	Council of European Municipalities and Regions
CIB	Construction Industry Board
CIBSE	Chartered Institute of Building Services Engineers
CIOB	Chartered Institute of Building
CIOH	Chartered Institute of Housing
CISC	Construction Industry Standing Conference
CITB	Construction Industry Training Board
COSLA	Convention of Scottish Local Authorities
CPRE	Council for the Protection of Rural England
CRE	Commission for Racial Equality
CSD	Committee on Spatial Development
CSO	Central Statistical Office
CVS	Community Volunteer Service
CWP	Constructive Women Professionals
DCPN	Development Control Policy Note
DDA	Disability Discrimination Act

DETR	Department of the Environment, Transport and the Regions
DG	Directorate-General
DLO	direct labour organisation
DoE	Department of the Environment
DoT	Department of Transport
EAGGF	European Agricultural Guidance and Guarantee Fund
EC	European Commission
EEA	European Environment Agency
EFTA	European Free Trade Area
EHTF	English Historic Towns Forum
EIA	Environmental Impact Assessment
EIS	European Information Service
EMU	European monetary union
ERDF	European Regional Development Fund
ESCTC	European Sustainable Cities and Towns Campaign
ESDP	European Spatial Development Perspective
ESF	European Social Fund
ESRC	Economic and Social Research Council
ETB	English Tourist Board
EU	European Union
FoE	Friends of the Earth
FOPA	Feminist Organisation of Planners and Architects
GDP	gross domestic product
GLC	Greater London Council
GNVQ	General National Vocational Qualification
HBF	House Builders' Federation
HFA	Health for All
HMSO	Her Majesty's Stationery Office
ICE	Institution of Civil Engineers
ICLEI	International Council for Local Environmental Initiatives
ISE	Institution of Structural Engineers
LA	local authority
LATI	Local Area Tourism Initiatives
LEEL	Lothian and Edinburgh Enterprise Ltd
LEOF	London Equal Opportuniteis Federation
LGIU	Local Government Information Unit
LGMB	Local Government Management Board
LPAC	London Planning Aid Consortium
LRN	London Regeneration Network
LWMT	London Women and Manual Trades
NDPB	non-departmental public body
NGO	non-governmental organisation
NPF	National Planning Forum
NVQ	National Vocational Qualification
OECD	Organisation for Economic Co-operation and Development

ONS	Office of National Statistics
OOPEC	Office of Official Publications of European Communities
OPCS	Office of Population Censuses and Surveys
PDD	Planning and Development Department
PPG	Planning Policy Guidance Note
RC	Rural Challenge
RDC	Rural Development Commission
RDP	Rural Development Programme
RIBA	Royal Institute of British Architects
RICS	Royal Institution of Chartered Surveyors
RNIB	Royal National Institute for the Blind
RTPI	Royal Town Planning Institute
SAPN	Scottish Anti-Poverty Network
SBC	Swindon Borough Council
SEA	Single European Act
SEF	Scottish Environmental Forum
SEM	Single European Market
SEPA	Scottish Environmental Protection Agency
SNF	Scottish Natural Heritage
SOBA	Society of Black Architects
SODD	Scottish Office Development Department
SRB	Single Regeneration Budget
TBC	Thamesmead Borough Council
TCM	town centre management
TDAP	Tourist Development Action Programme
TEC	Training and Enterprise Council
UN	United Nations
UNCED	United Nations Conference on Environment and Development
UTO	United Towns Organisation
VMP	Visitor Management Plan
WAC	Women Architects' Committee
WCC	Women's Communication Centre
WCED	World Commission on Environment and Development
WDS	Women's Design Service
WGSG	Women and Geography Study Group
WHO	World Health Organisation
WWF	World Wildlife Fund

Part I
Context

1 Introducing social town planning

Clara H. Greed

Planning revisited

The purpose of this book is to investigate 'social planning' with reference chiefly to the 'social aspects of planning' agenda within British town planning. This, inevitably, leads into a consideration of issues such as urban governance, social policy, inequality, sustainability and planning theory. This book develops further the theme of investigating the social construction of urban realities and thus of social town planning which was introduced in earlier work by the editor (Greed, 1994) and which has long intrigued the various contributors (for example, R.H. Williams, 1975, *inter alia*). Nowadays it may be argued that there is not one 'town planning' but many new 'plannings', each with its own agenda, devotees, and priorities, including, for example, environmental planning; urban design planning, Euro-planning; and market-led urban renewal planning (cf. Greed, 1996a; Greed and Roberts, 1998). One of the most dynamic, changing and controversial of the 'plannings' is what may be broadly termed 'social town planning'. There has been a proliferation of demands and policy proposals to meet the needs of minority interests and community groups, which the present scope and nature of statutory town planning appears unable, and ill-equipped, to meet.

This book is aimed at good second year students and above who are studying town planning, urban geography and social policy, whilst providing debate for more advanced readers. Those readers new to the subject who require historical background are referred to chapter 11, 'Urban social perspectives on planning' in *Introducing Town Planning* (Greed, 1996b), which provides an introductory summary of the issues, and to chapters 7–9 of *Women and Planning* (Greed, 1994) which deals with the issues in greater depth. In a nutshell, much of the debate over the role of planning over the last 150 years has revolved around the question of whether it is possible to achieve 'salvation by bricks': that is, whether social problems can be 'solved' by redesigning the built environment (a reformist viewpoint), or whether such efforts are simply useless, 'like rearranging the deckchairs on the *Titanic*' (a more socialist view), because, it is held, more radical change is needed within society itself before cities will change. 'Environmental determinism' is the term used, often pejoratively, to describe the planner's attempts to play God (cf. Newman, 1973; A. Coleman, 1985). In contrast, nowadays one is more likely to find that 'space' (the physical layout) is seen as a secondary consideration in some branches of social town planning, and greater emphasis is put upon 'aspatial' (social) urban policy in respect of health, sustainability, inner-city regeneration, and community empowerment. But many would still argue that 'space matters' (McDowell, 1997), especially for women seeking to negotiate the city of man, and the disabled whose urban mobility and access is limited by unnecessary steps and poor design.

DEFINITIONS AND CONNOTATIONS

Social town planning and social planners

But what is social town planning? And who are the social town planners? Arguably, there is no one definitive answer to the question 'what is social town planning?' for relatively speaking, as with all town planning, 'it can be anything you want it to be' (as discussed in Greed, 1996b, ch.1).

'Social town planning' may broadly be defined as any movement to introduce policies that take into account more fully the needs of the diversity of human beings who live in our towns and cities, (which many would argue mainstream town planning has failed to do). The need for 'putting the social back into planning' must be seen within the context of a statutory town planning system in Britain that was set up to deal with physical rather than social issues. Typically, emphasis is put upon 'land-use' planning, primarily as reflected in land-use zoning and the creation of spatially focused development plans. Likewise, town planning law has been obsessed with proving 'change of [land] use' rather than facilitating the way in which people 'use land'. The crux of the book pivots around this physical/social dualism, and all the various contributions are united by an underlying consideration of the unequal spatial/aspatial balance within existing town planning.

This physical/social dualism has been particularly highlighted by the 'women and planning' movement, which has sought to introduce more gender-sensitive approaches to town planning (Greed, 1994; WGSG, 1997). It has been demonstrated that the planning process is not 'neutral' but tends to benefit those groups dominant in society itself. Attempts to include 'social' policies on childcare, for example, in a development plan are likely to be met with the DETR response that the policy in question is *ultra vires*, because it is 'not a land-use matter'. But, as will be seen, a whole range of other agendas and issues have also fallen foul of the *ultra vires* trap, especially in respect of policies that reflect the needs of ethnic-minority groups, the elderly and disabled. An upsurge of concern with 'equality', 'minorities', and related 'isms' over the last twenty years has moved/forced town planners to accept a wider range of social considerations when 'doing town planning'.

Many commentators consider that town planners, along with other built environment professionals, have been somewhat reluctant and slow in their response to minority issues (as commented upon by the various contributors). Much of the initiative for change has come not from the planners themselves, but from community and minority pressure groups at grass-roots level. In spite of the growth in publications, publicity and conference events in this field, there is little evidence that planners are 'walking the talk'. New legislation, PPGs (DoE/DETR Planning Policy Guidance Notes), and regulations have not been rushed through to meet the town planning demands of women, ethnic minorities and the disabled, and one seldom finds more than a short, 'obligatory' token mention of the needs of these groups in central and local government planning documentation (cf. DoE, 1997a).

This half-hearted response to 'minority' needs contrasts markedly with earlier attempts to undertake social town planning in relation to meeting the presumed needs of the 'working classes'. For many years 'class' alone appeared to be the main consideration and organising concept in delineating the 'social aspects of planning'. This was epitomised in the more socialist times of the post-war reconstruction period in the emphasis upon 'planning for the working class', which served to legitimate the questionable nature of much planning policy, such as slum-clearance, New Town building and regional policy. Significantly, no one cried '*ultra vires*' (or said, 'class – that's not a land-use matter') when huge chunks of our inner cities were demolished and their inhabitants put into neighbourhood units in new town developments planned to generate 'community spirit', but many complained about the abuse of political power, legitimated by academic social theory, in such urban pillage (Dennis, 1972).

Town planning has always contained a mixture of components, some of which, such as economic and political elements, cannot be categorised as purely 'social' or indeed 'physical' in nature. Planning is a political process, informed by a range of ideologies (cf. Kirk, 1980; Healey, 1991; Montgomery and Thornley, 1990). Urban social theory, which influences social planning, is seldom neutral. Planning is inevitably 'political' because it is concerned with land and property, and the allocation of scarce resources. It has therefore been scrutinised by those concerned with understanding 'capitalism', and class and power structures within society. Social town planning has frequently been associated with 'the Left', be it in the policies of the Greater London Council at its height (GLC, 1984) or as part of full-blown Eastern European state socialism, but 'planning' has attracted interest from those right across the political spectrum. Second, at the implementatory level, the planning process is a political activity because it is concerned with the allocation of scarce resources, the planner being seen as acting as urban manager or social policy-maker (Simmie, 1974). Third, the planning process is extremely political at the local-area-plan level, where community politics and grass-roots activity thrives, as will be illustrated by various case studies.

Who, though, are the 'social planners'? In the past many 'urban social problems' seemed suitable specifically for treatment by *town* planners, because it was imagined that one could draw a line around the 'bad area' and bombard it with physical planning remedies. However, as will be seen, many of the social issues of concern today are not spatially confined to one area. Whilst so-called ethnic minority areas might be physically identifiable, gender and disability issues occur everywhere and cannot be contained by 'special area' policy. For example, access policy requires a city-wide approach to implementation, wheresoever barriers to mobility exist across the whole built environment and within the attitudes of the population as a whole. Likewise, the increasing concern with environmental sustainability cannot be spatially limited but socially affects everyone on the planet. Nowadays there are a range of other 'social' planning issues that affect the whole city, which are not necessarily associated with deprivation or 'social problems' but rather with greater affluence, such as the growth of tourism, the development of the arts and 'culture' and millennium project planning. Thus a whole range of different types of policy-makers and professionals are involved in 'planning' issues, many of whom would not see themselves as town planners. Because 'planning for sustainability' is such a prominent issue in many of the chapters of this book, as it has become increasingly linked with 'social', rather than purely 'environmental' issues, there follows a short explanation of the term 'sustainability'.

Sustainability

The influence of the sustainability movement on social town planning merits a digression because it is a concept that informs the approach taken by many of the contributors. Sustainability is a key consideration at all levels of social planning nowadays. At first sight, 'sustainability' in Britain at least, may be associated primarily with environmental planning policy, and the relationship between the 'social' and 'environmental' agendas often appears unresolved and ambiguous at best, totally separate at worst. However, internationally, it may be defined as having four elements: to conserve the stock of natural assets; avoid damaging the regenerative capacity of ecosystems; achieve greater social and economic equality and equity; and avoid imposing risks and costs on future generations (Blowers, 1993). Equity may be seen as a further development of the concept of 'equality' pioneered by minority groups, and includes the idea of equality both between people, and within human relationships and dealings, not least within the political and economic context of world development. Such equity planning clearly overlaps with concepts of social town planning, but combines a strong spatial component. These principles were embodied in the Bruntland Report (WCED, 1987) and in the Earth Summit in Rio, in 1992. Following Rio, the Agenda 21 programme was established, requiring all signatory states to draw up national plans on sustainability. In Britain, the Department of the Environment produced in 1994 the policy document *Sustainable Development: The UK Strategy* (DoE, 1994a). Likewise there have been a series of EC (European Commission) Directives increasing environmental controls within the European member states, and a range of federal and state-level measures introduced in North America. However, in some north European countries (not least Sweden; Guinchard, 1997) environmental impact analysis has also been strongly linked to the measurement of the social impact of planning policy, and greater emphasis is given to equity, especially gender equality, as an integral component of sustainability (Skjerve, 1993).

Environmental assessment and environmental impact studies are now commonplace in most developed countries in respect of evaluating the implications of large-scale urban or rural development. Chapter 28 of Agenda 21 required local government bodies in each member-state across the world to produce 'Local Agenda 21' initiatives which translated the overall objectives (UNCED, 1992). Attention has been given to formulating urban development policies which seek to reduce pollution, reducing use of fossil fuels and other natural resources, in conjunction with consideration of the health implications of such policies and the accessibility dimensions. As a result, town planning policies based on meeting the needs of the car are being replaced by a greater emphasis upon control of the car, encouragement of public transport and reducing the need for people to travel in the first place. Policy measures include promoting urban containment, discouraging decentralisation and dispersal of land uses and facilities and encouraging higher density urban districts which contain a balanced diversity of facilities, amenities, and land uses, so enabling greater local economic and social sufficiency, and thus urban sustainability.

Clearly, many of these issues have social implications and dimensions, some planned and some unforeseen. Whilst welcoming the greater emphasis upon nurturing the environment, some among minority groups are wary of restrictive transport policies in situations where, because of the decentralisation, demolition and destruction sanctioned by previous generations of planners, the only shops for essential food shopping are now out of town hypermarkets that require the use of a private car or a long, arduous journey by infrequent public transport. Now that cities have been replanned around the car, many are concerned about the social implications of putting the clock back *before* implementing major relocational policies for retail, school, and employment locations. Out-of-town clearance on the scale of 1950s inner urban area clearance and relocation, through market forces or state intervention, may yet prove the logical solution. Thus the old 'red' (socialist) agenda of much social town planning has increasingly become replaced with a new 'green' (environmental) agenda that challenges not only social inequality but also personal lifestyles and urban culture itself.

Contents

Style and components

'Social town planning' is a potentially enormous field, without clear boundaries or demarcations, and key elements, such as 'women and planning', do not fit into existing statutory planning compartments. In these days of post-structural 'diversity' one feels freer to mix different components from a diverse range of realms than was possible under traditional taxonomy. This book seeks to capture key aspects of the diverse and evolving manifestations of social town planning. The contributions are presented in a variety of styles. Because some of the issues discussed have affected individual contributors in their private and professional lives, a personal style is used in some of the chapters. Contributors come from a wide range of academic realms and professional fields. Depending on its appropriateness to the subject being presented, some contributors include policy guidance in their chapters, whereas others concentrate upon analysis and the development of conceptual perspectives. This mixture reflects the diversity of the discourse of social town plannning.

In summary, Part I provides background on the nature of social town planning from both a historical and a current professional planning perspective. Part II investigates a representative selection of (so-called) minority issues and (what are popularly seen as) social planning topics, including gender, race and age, but cross-links these components with mainstream planning issues such as housing, rural planning and transport. Whilst Part II looks at individual topics at the 'micro' level (albeit with 'macro-level' city-wide policy implications), Part III, in contrast, takes a wider perspective, discussing a range of global and European organisations and initiatives which are, in effect, promoting social town planning objectives. These overarch British town planning, and may offer means of escaping and resolving the spatial/social impasse. Part IV initially steps back from the spatial/social debate and looks at 'the problem' from a wider cultural perspective. It is argued that cities are not just the result of the 'reproduction over space of social [and economic] relations' (Massey, 1984, p. 16), but that they are the ultimate cultural artefact: they are the outward manifestation of inwardly held beliefs about what 'reality should be like'. The organisation of time, space and activities within cities arguably owes more to cultural beliefs, tradition and taboo than to the efforts of modern planners (particularly in respect of gendered spatial divisions: see Domingo and Lopez, 1995). The final chapter addresses the question of how to change the culture of professional planners, in order to transcend the physical/social dualism and transform social town planning.

Geographically, the book is centred on Britain; however the book is not just 'English' – there is a strong emphasis upon Scotland. Additional insights and examples are drawn from Europe, Australia and North America. Rural, as well as urban, planning topics are covered in this book, but many social planning issues are supra-urban and cannot be taken in spatial isolation. Within British town planning (whose full statutory title, 'town and country planning', is often abbreviated to the generic 'town planning') planners are constantly concerned with the relationship between town and country, and especially with urban containment within limited land space, in which even rural settlements and their populations are effectively part of urban society and a predominantly urban national culture. Indeed, the urban/rural distinction blurs when one is taking a global sustainability perspective.

Part I: context and background

Following this chapter, rather than providing, yet again, a historical account of the development of town planning and the related evolution of social (and socialist) concerns within it in chapter 2 Huw Thomas moves instead straight into a discussion of the validity of town planning containing a 'social' component, with reference to the role and priorities of professional town planners, in order to explain the nature of 'the problem'. Whilst chapter 2 illustrates the implications of the social/physical divide within planning on the mindset of planning education and practice, in chapter 3 Nigel Taylor reflects upon the historical development of town planning theory in the post-war period. Because the work of Lewis Keeble, the ultimate physical land-use planner, is so seminal to the development of modern town planning, Nigel takes as his starting point an interrogation of a quotation from his *Principles and Practice of Town and Country Planning* (1952).

He shows that Keeble's work provided a 'bible', an authoritative reference providing clear-cut answers for post-war town planners. But town planning soon proved to be not so straightforward, because it was realised that society is composed of numerous different groups, holding different values and interests with respect to the kinds of environments they would like to see, from which it is now widely inferred that there are no simple, or universal `principles' of town planning. Indeed, post-Thatcherism, the very existence of town planning practice as an institutionalised activity of the state is contested: in retrospect, the writings of Lewis Keeble now strike us as naive. Yet Nigel concludes that what Keeble wrote in the piece quoted at the head of chapter 3 is at least partly correct and, in this part, importantly correct. Keeble's view of town (and country) planning contains a useful corrective to those planning theorists who imagine that town planning has the remit to do all kinds of 'social and economic' things, or that town planning is 'social and economic' planning as distinct from just plain physical planning.

In the second and main, part of the chapter he subjects the social/physical 'question' to logical analysis, in the form of a debate as to what it actually means to talk of town planning being 'physical' and/or 'social'. This structured interrogation yields a set of intellectual tools, including propositions and arguments, that might profitably be utilised to evaluate the validity of the assertions of current manifestations of social town planning as presented by other contributors to this book.

Part II: groups and issues

This section explores a range of so-called minority issues which are popularly seen to be part of the equal opportunities agenda, and which in many planners' minds appear to equate with what they imagine 'social town planning' to be: namely planning for the 'other'. In contrast to this, each contributor here implies that their topic affects everyone and requires a fully inclusive approach. Many of the chapters are 'double-barrelled', as they deal with a minority social group in relation to a particular spatial policy area. Most contributors assume prior elementary knowledge of the issues in question, but references are provided to both basic and more advanced material for those unfamiliar with the field.

In chapter 4, Jo Little discusses gender issues, but this time from a more rural perspective within the context of doing social 'town' planning. Despite some recognition by academics and policy-makers of the problems experienced by women living in rural areas, planning initiatives remain virtually non-existent.

Using examples, she documents the lack of sensitivity towards issues of gender inequality within rural planning policy. In contrast, she presents examples of where women, working independently, have instigated change at the community level through the introduction of, for example, childcare initiatives and employment co-operatives. Reference is made to current theoretical debates. It was considered worthwhile to undertake a specific study of the rural situation, as against an overview of all aspects of 'women and planning', as the former provides an opportunity to combine a social issue with a specific physical, spatial planning setting and also enables her to show how her research links to planning practice.

Chapter 5, by Rose Gilroy, examines the way planning, which is taken to include town planning, community care and local authority service provision, is responding and could respond, to the changing position and number of older people in Britain. The chapter raises some interesting questions about the peculiarly limited scope and nature of British town planning. Those over pensionable age now make up nearly a fifth of the population, although, arguably, the elderly are also increasing their power politically as well as demographically. What quality of life do the elderly have in the community? Local authority house-building programmes have been cut to practically nothing, while the Housing Corporation has also been hit to the detriment of housing for elderly people. Local authority services for older people, provided through social services or grant-aided through the voluntary sector, have also been cut particularly hard. People can no longer take it for granted that the state will look after them, as the Welfare State and the concept of cradle to grave provision is now seen as an out-dated concept. These social policy changes and related negative images of the elderly as a 'burden' are discussed. Town planners' approaches to planning for the elderly are examined in the light of cultural and conceptual images of old age within society itself. The chapter also examines issues of poverty, access, fear of crime and the aspirations of older people themselves. Elderly people's expectations of the town planning system and the role of participative forms of planning consultation in eliciting their views, are discussed, drawing on research undertaken by the contributor.

In chapter 6, Linda Davies looks at the question of planning for disability (or rather more positively, creating 'barrier-free living'). Planning for disability has been the Cinderella of the planning process. In contrast to the great emphasis that has been put upon social class and regional differences in developing planning policy, disability policy and legislation has generally been developed outside the town planning system. Linda discusses definitions of disability, reviews legislation and regulation and provides a series of case studies. A key theme is the move away from a mentality of physical control (through the Building Regulations) towards enabling social access (through planning policy). This chapter contains the most detailed, specific policy material of all the chapters, centred upon the design of the frontage of one high street bank. Such, though, is the nature of this topic, in which social and physical factors are very closely linked. The case studies demonstrate how very detailed 'little' physical design issues (such as just one doorstep) and 'small details' (such as aspects of staff training) can have macro-level social implications, preventing or enabling equal and unhindered access to the built environment for the disabled, and thus, effective social town planning.

Chapter 7, by Geoff Vigar, explores the social implications of currrent UK transport policy. The editor strongly wanted a chapter on transport in this book, because although it is a planning specialism renowned for its extreme male domination, unsociological agenda and 'techno-fix' mentality, transport policies have immense social implications. For example, some central government surveys have traditionally not counted journeys by foot and journeys of under one mile, thus excluding many people and their essential trips from the data (see ONS, 1996, Table 14.1 and annually). Thus this chapter deals with an 'issue' rather than a 'minority' per se, but it soon becomes evident that the 'carless' are the hidden and unmentioned minority in the transport discourse and that this group overlaps with the different minority groups discussed in other chapters. The chapter first discusses the nature of transport policy and then reflects upon mobility trends, suggesting there is a mismatch. The main part of the chapter explores the concepts of accessibility, mobility and the potential for principles of social exclusion or inclusion to put a social agenda back into transport planning. Recommendations are made for creating transport policy that is more inclusive of social issues.

In chapter 8, the issues of social inclusion or exclusion in respect of minorities, experiences of the world of town planning and architecture are illustrated through a personal account by Ann de Graft-Johnson, a Black woman architect who brings into the discussion the three issues of race, cultural diversity and gender. The issues are considered at a series of levels, from the international context 'down' to the local British situation. Through her professional experience, personal witness and academic research, the author looks at how the social and physical aspects of designing are interconnected. Her perspective is that of an architect whose concerns are, on the one hand, the need to ensure close links between practice and theory and, on the other, that mechanisms for review and assessment contribute to the healthy evolution of a cultural inclusive environment, thus transcending the social/physical dualism.

Part III: new policy horizons

This section presents and discusses changing agendas, perspectives and policy directions, first from Europe and then from the wider international setting, which might influence the UK approach to town planning and enable a more socially inclusive approach to policy-making and practice. The interrelationships found in other systems between environmental, health, social and economic planning policy are highlighted and the implications for British conceptualisations of 'physical' land-use planning are reflected upon. The European and global perspectives and ideas presented provide a source for inspiration to those wishing to break the spatial/social deadlock. In parallel, many women planners are encouraged by pan-European networks such as Eurofem, which promotes women and planning issues within the nation states and at the EU level (Eurofem, 1998).

The central theme of chapter 9, by Richard Williams, on the European Union, is how far the social planning agenda in the UK, as defined and discussed elsewhere, is shaped or framed by the EU context. The starting point is the key phrase 'economic and social cohesion', which occurs in many EU contexts and the meaning of the word 'social' in this phrase. Economic and social cohesion can refer to anything from cohesion of the EU as a whole to cohesion within cities with problem neighbourhoods. The chapter reviews the EU competences and sectors of EU policy which address, or are intended to address, the question of social cohesion and explores the ways in which they seek to meet social planning objectives. Sectors include the Social Chapter, the European Social Fund, Community Instruments and various urban initiatives. Spatial policy and the European Spatial Development Perspective, which seeks to provide coherence to policies designed to promote economic and social cohesion, are also examined. Links are made between the spatial scales, from the supranational through the transnational and interregional to the urban, drawing upon the author's experience of programmes such as INTER-REG. The quality of urban life is also considered.

Chapter 10, by Colin Fudge, explores urban change within the context of rapid global urbanisation, including the emerging urban dimension to European policy and the growing interest in urban sustainable futures. As the chapter develops, the links between European and international initiatives are delineated, moving the reader up one level, to an awareness of the global dimensions of planning. Within this context, the nature of urban health and the development of the WHO healthy cities programme are discussed. This initiative is linked to the parallel development of the European Sustainable Cities Project and the actions of the EU Expert Group on the Urban Environment and the European Sustainable Cities and Towns Campaign. A concluding section examines the influence of these initiatives on European urban policy for cities in the twenty-first century. In 1996 the point was reached at which over half the world's population now live in cities many in overcrowded, poor and insanitary conditions: thus 'health planning' inevitably becomes a component of social town planning.

Chapter 11, by Janet Brand, 'Planning for health, sustainability and equity in Scotland', illustrates how the new international agenda stemming from the Rio Declaration and Agenda 21 may assist to resolve the social/spatial dualism inhibiting the fulfilment of social planning within the current UK statutory planning system. She outlines the key components of sustainability, which are defined as social equity, economic self-sufficency and environmental balance. Focusing upon the equity component and participatory models, the Local Agenda 21 process and its operation in Scotland, the chapter is illustrated by case studies in which multi-agency initiatives are making linkages between planning, health, sustainability and equity. This chapter shows how global and European initiatives (identified in chapters 9 and 10) can be used at the local authority level in respect of specific planning projects. The Scottish context provides fresh perspectives on resolving the physical/spatial dualism and suggests alternative means of 'doing planning' which incorporate equity, sustainability and health within a physical land use planning context.

Part IV: cultural perspectives on planning

In the final section of the book, social town planning is approached from a cultural perspective, as a means of transcending the physical/social dualism by using alternative agendas and approaches to planning. In chapter 12, Franco Bianchini, informed by comparisons with European examples of cultural planning, reflects upon the nature of British town planning. He argues that a great weakness in the British system, resulting in ugliness of urban design, has been the lack of emphasis upon 'art' and 'culture' in the planner's training and the separation of town planning education and practice from a concern with urban arts, social activities and cultural activities such as music, dance, gastronomy, ritual and tradition. He argues for a more inclusive approach in which town planning is closely linked to arts and cultural policy and activities within cities. An additional section by the editor summarises another approach to planning, originating in Italy, namely 'time planning', which developed out of a more culturally sensitive approach to town planning. In this, 'time' rather than 'space' is taken as the main organising factor in achieving socially satisfactory town planning.

Nowadays, there is arguably greater emphasis being put upon arts and culture within urban policymaking, not least because of the increased funding being made available from the Arts Council, the Lottery and millennium projects. There still seems to be greater emphasis upon individual 'public art' projects and schemes rather than upon seeing all urban policy-making as a form of urban design and art (Greed and Roberts, 1998, pp. 199–200). However, it is in the interests of local authorities to improve their cultural profile and to protect and conserve their urban historical heritage, if only to gain increased tourist revenue. In the course of developing this book, the editor found new courses increasingly being developed in town planning and geography departments, which offered degrees in 'tourism' that still included a substantial amount of 'town planning'. In some colleges it appears that erstwhile town planning courses have simply been renamed 'tourism studies' to cash in on the enthusiasm among young people for degrees in arts, media and leisure studies and to counter the decline in popularity of built environment degrees. It might be argued that the field of tourism studies comprises a new form of social town planning concerned with planning for large numbers of people who are affluent and pleasure-seeking (not poor and deprived) within specific urban areas and which has links with urban conservation, urban heritage and cultural policy. It was therefore decided to include a chapter on 'planning for tourism'. In chapter 13 Rob Davidson and Robert Maitland argue that tourism policy is an integral part of town planning which links directly to local economic and social planning, within the context of the entrepreneurial, proactive format that town planning nowadays takes.

In chapter 14, the relationship between 'culture' and 'social town planning' is investigated further by Jean Hillier, who considers the cultural and community factors involved in achieving effective communication in the planning process, with particular reference, for comparative purposes, to planning 'with' Aboriginal people in Western Australia. This chapter provides a reflective consideration of the role of cultural perceptions and images in planning policy-making and procedurally just communicative action in working towards social justice. Whilst Janet Brand's chapter stresses the triple top of equality, sustainability and health, Hillier's chapter juggles with the trio of equality, environmental sustainability and cultural justice for minorities as key components in sensitive social town planning. Hillier, for example, is concerned that the cultural differences exposed during interactive discourse lead to planners and planned 'talking past' rather than negotiating with each other (a common complaint within the inner city in Britain too). This chapter is concerned with planning processes in relation to participatory, procedural and implementatory issues: that is, about how to 'do social town planning' in such a way that the social outcomes are reflective of the needs of the community and create greater equality.

A common thread through many chapters is the importance of the professional subcultural values held by town planners as to 'what is normal', in informing their approach to policy-making and in their estimation of fellow professionals' worth. In the final chapter the issue of 'culture' is revisited as a possible key to unlocking the physical/social conflict inherent within the planning system. Drawing on recent ESRC-funded research on the changing composition and culture of the construction industry and of the built environment professions (Greed, 1998) key change agents, including minority groups, that might bring about change within the world of social town planning are identified.

2 Social town planning and the planning profession

Huw Thomas

Professional and organisational perspectives

This chapter reviews, and tries to explain, the fluctuating (but generally marginal) interest taken in social welfare issues (community development, race, social justice, etc.) by the planning system since the 1900s, paying particular attention to the post-war period and the present day. Central to the analysis will be a discussion of the growth of the planning profession itself. Occupational closure exercised by the dominant professional subcultural group has had a profound effect on how the roles, priorities and concerns of planning have been defined. Town planning operates within the constraints of local government structures, and the significance of professionalism and departmentalism in local government will be discussed. It will also be suggested that professional planners have been willing to, or felt obliged to, trim their views of the relevance of a concern for social welfare to planning according to, firstly, local political circumstances, and secondly, the legal framework within which planning operates. As well as analysing professional documents and activities, therefore, the chapter will also refer to judicial interpretations of the proper score of planning. The chapter concludes that a plausible explanation for the weakness of social town planning as a professional goal must be based on the idea of planning as a state activity intended to manage, not necessarily benefit, the worse-off members of society.

In addition to reviewing previously published work (such as Healey on typologies of planning styles and Harrison on social welfare and planning), the chapter draws upon the author's own research into local governance and planning. The ways in which professional planners in post-war Britain have defined their tasks are examined and, especially, their views of how these tasks should relate to a concern for the welfare of the populations of the settlements they have planned (and the distribution of welfare – of costs and benefits – among particular groups within these populations).

It would be mistaken to assume that any profession is homogeneous and that at any time all its members share one set of views or a single professional ideology (Bucher and Strauss, 1961). On the contrary, professionals working in different settings and hence subject to different pressures and career opportunities

Table 2.1 Healey's typologies of role models for planners

Role	Influence in British planning
Urban Development Manager Planning as 'the production and management of good urban design and urban development'	'the main role model . . . until the 1970s'
Public bureaucrat 'performing duties defined by politicians'	In practice, very influential from the 1940s onwards
Policy analyst 'the planner is a policy scientist'	Limited influence from the 1970s
Intermediator 'deploying interpersonal skills in negotiating and social learning'; planner as implementor	Influential from the late 1970s onwards
Social reformer 'committed to changing society'	A persistent strand in planning ideology

Source: Healey (1991), p. 14.

might to be expected to develop different perceptions of their own roles and of what constitutes the core of the professional task. Moreover, if occupational settings change over time, if the pressures on and expectations of, professionals change, then we would expect their view of their profession to be affected. In relation to planning, Faludi (1972) has argued very persuasively that there have been systematic differences in professional ideology between planners in consultancy and those employed in local government.

Nevertheless, as will be pointed out later in this chapter and as Faludi has himself argued, post-war planning in Britain has been dominated by employment in local government. Though local government is not itself homogeneous, there has been enough commonality of experience, it is suggested, to allow typologies of dominant or favoured professional approaches to planning to be constructed. Healey (1991, 1997) has suggested that it is possible to discern a succession of professional roles that have been widely held, or were perhaps important ideals or aspirations within the profession, at various times since the early twentieth century (see Table 2.1).

Healey's typology provides a chronological (and admittedly generalised) account both of the changing self-perceptions of some planners and of the debates among professionals and academics. Of the five 'role models' she identifies, the first three (and especially the first two) have been the most significant in British planning practice; yet, among these a wide range of evidence suggests that consideration of the distribution of social welfare is marginal or absent. The remainder of this chapter will sketch an explanation of why this might be so.

Social town planning at the margins

In the twentieth century, British town planning has been a state sponsored activity, conducted within a framework of law (including, in the Anglo Saxon tradition, case law) and largely undertaken by employees of central and – especially – local government. Town planners have been quite successful in establishing a degree of *de facto* social closure (Larson, 1977) in relation to their profession. Though it is not illegal to practise planning without planning qualifications, it is unlawful to term oneself a chartered town planner unless one is a member of the Royal Town Planning Institute (RTPI), entry to which is by relevant qualification and experience: a classic 'qualifying association' model of professionalism (Millerson, 1964). Until the early 1990s, the RTPI had its own set of qualifying examinations but these ran in parallel with another route to entry, namely by gaining a university degree which the Institute recognised (or accredited) as allowing exemption from its own exams. The latter route became increasingly popular as higher education expanded and has long been the main route into the profession.

More importantly, membership of the RTPI (or – in the case of less experienced planners – eligibility for membership) is widely regarded as an extremely useful qualification for employment, especially within local government (where for over thirty years no fewer than 70 per cent of planners have been employed). A national survey of local planning authorities conducted in 1991 illustrated the significance of 'credentialism' in planning: it found over 75 per cent of professional staff either had or were studying for a planning qualification (and qualifications, such as degrees in planning, generally pave the way to membership of the RTPI) (LGMB/RTPI, 1992, p. 26). The significance of professionalism is further borne out by a recent national survey of RTPI members. While only a third of respondents said that membership of the Institute was essential for gaining their current job, over 60 per cent gave as reasons for becoming members of the RTPI some kind of material self-interest (e.g. career progression, salary-scale progression or professional credibility) (RTPI, 1997).

This brief résumé of the institutional and legal context of the planning profession suggests that it is not unreasonable to regard the nature and scope of planning practice (and its relation to social welfare) as constituted by practices in five areas, namely:

- the education of planners;
- activities and debates within the professional institute;
- the departmental context of town planning, especially within local government;
- the judgements informing and embedded within the decisions and products of the planning system;
- the legal framework and judicial interpretation of it.

This section will do no more than provide some examples from each of these areas; they will not amount to a comprehensive description of how social welfare has been regarded in planning, but the consistency of the direction in which they point is extremely persuasive.

Consider, first, the education of town planners. When part-time study and so-called direct entry were major routes into the RTPI it is reasonable to suppose that practically oriented textbooks such as those of Lewis Keeble (1952) (as discussed further in chapter 3) were extremely influential, as they provided a kind of 'one-stop shop' for students with limited time for study. Throughout a long career as practitioner, academic and activist within the RTPI, Keeble seems to have stuck to a view of planning which:

1   defined the specific expertise of planners as the ordering of the use of land and buildings;
2   asserted that successful planning of this kind improved social welfare in general;
3   refused to accept that planners should try to shape social behaviour, a view which slid into a refusal to consider the distribution of the costs and benefits of specific planning decisions.

These points are illustrated here by quotations from the second (1959) edition of his textbook, though similar (often identical) quotations could be found in the other editions. So, town and country planning is:

> the art and science of ordering the use of land and the character and siting of buildings and communication routes so as to secure the maximum practicable degree of economy, convenience and beauty.
>
> (Keeble, 1959, p. 1)

And its contribution to aggregate social welfare is clear:

> socially successful planning tends to make people's lives happier because it results in a physical environment which conduces to health, which allows convenient and safe passage from place to place, which facilitates social intercourse and which has visual attractiveness.
>
> (ibid., 1959, p. 9)

This faith in physical planning as a (socially) good thing appears to have been widespread in the 1940s, 50s and 60s. Thus McAllister and McAllister's polemic in favour of comprehensive town planning states:

> Instead of town chaos we see town order. Instead of conditions which breed disease, unhappiness and crime, we see a physical environment making for health, happiness and a positive social life.
>
> (McAllister and McAllister, 1941, pp. xxv–xxvi)

Yet though the tone of this quotation would undoubtedly have been congenial to Keeble, its final evaluative reference would have left him anxious. He argued for what he termed 'social neutrality' (Keeble, 1959, p. 9) in planning and warned planning authorities against discriminating 'between land uses involving activities of which they did not approve' (p. 10). Carrying this idea to what might seem extreme lengths, he claimed that the idea of the 'neighbourhood', so central to post-war planning (especially in New Towns) should properly be understood as denoting a *service centre* and should not be imbued with some wider social significance (pp. 162–163). Planning, in this view, was not about creating socially mixed communities (p. 163) or the distribution of wealth (p. 9). It is undoubtedly the case that, Keeble notwithstanding, the idea of social mix has existed within debates about planning, but the naivety (and perhaps lack of interest) of planners in dealing with economic and social concerns is nowhere better demonstrated than in the more or less complete absence of any sustained, planned mixed class/mixed tenure neighbourhoods in the UK.

Not surprisingly, Keeble regarded the contribution of social science to planning as a very limited one – a kind of market research (see Keeble, 1959, pp. 105–6). By the late 1960s there were textbooks that defined a more significant role for social science. For example, McLoughlin's (1969) conceptualisation of physical planning as 'the control of complex systems' required that planners had an understanding of the dynamics of the so-called 'components' of the system (human activities) and their 'connections' (human communication) (ibid., pp. 77–8). This kind of understanding was, in essence, social scientific, but McLoughlin pointed out that its purpose was to assist the planners in their specialised role of shaping the physical form that such systems took. In principle, there was no reason why the system should not be guided towards egalitarian social welfare goals; in practice, enthusiasts for the approach focused on the complexity of understanding urban systems rather than the complexities of debating issues of social justice. As Donnison (1975, p. 267) put it:

> the whole system's literature [has] practically nothing to say about what sort of a city or county we actually want or how to get it. Some authors thought they had only to formulate models which in a mathematical sense explain (i.e. represent) the process of urban development, without exploring its causes or evaluating its effects.

It is indicative of the ferment and fragmentation in planning education and planning thought from the 1970s to the present that Cullingworth and Nadin's *Town and Country Planning in the UK*, the most widely used textbook of the period, concentrated on descriptions of institutions and legal procedures rather than prescriptions for how planning should be undertaken. First published in 1964, it is now in its twelfth edition (Cullingworth and Nadin, 1997). In this latest edition, pages 35 to 280 are devoted to describing the agencies of planning, the roles of development plans, the mechanisms of development control and topics such as heritage planning, transport planning and urban policies (each with a heavy emphasis on the legal and institutional frameworks for policy-making). Pages 283 to 286 consider 'race and planning', 'planning and gender' and 'planning and people with disabilities'.

Within higher education there has been a reaction to the systems approach, which has included a concern for the distributive consequence of town planning. The contours of the debates have been mapped elsewhere (e.g. Healey, 1982; Hague, 1984 and, most recently, Sandercock, 1998) and for our purposes it suffices to point out that it is reasonable to suppose that the concern of academics for the social welfare implications of town planning must have had some influence over the curriculum of planning schools. Unfortunately, there is no systematic research of recent planning education, but research about its content in the 1970s suggests the co-existence of a variety of approaches, with social welfare continuing to be marginal in virtually all schools (Thomas, 1980). Three factors might reduce our surprise at the apparent paradox of 'cutting-edge' academic planning literature being concerned about social welfare, while the day-to-day education of planners is not. First, those contributing to the literature are a small minority of planning academics and may therefore be unrepresentative of their colleagues. Second, planning schools (and their students) are conscious of providing vocational courses. This leads to a call for 'relevance' in the curriculum, a criterion often defined narrowly as skills and knowledge that will help the graduate planner secure employment on leaving planning school. If 'social town planning' makes no inroads in professional practice, educators of planners will always be under pressure to minimise the space accorded it in the curriculum. The final significant factor is the guidance of the Royal Town Planning Institute (RTPI), which accredits degree courses in town planning. Hague (1996, p. 7) has argued that in the 1980s the RTPI's guidelines for planning schools shifted 'the agenda [of education] towards a form of town planning that was conscious of social and environmental impacts'. A reading of the Institute's guidelines over the last twenty years bears out the contention that there has been a real and worthwhile shift, but also qualifies enthusiasm about its significance.

In 1976, planning was defined in the guidelines as a spatial design ability which would – *inter alia* – assist in satisfying social and economic needs through regulating the physical environment (RTPI, 1976, p. 1). But in seventeen pages of guidelines, which included injunctions to understand social and economic systems, the only apparent reference to the distributional consequences of planning was in the two lines:

> Consideration should be given to the implications for the professional planner of . . . the ethical, political, social economic and scientific problems of modern society.
>
> (ibid., p. 6)

By 1991, awareness of the 'value dimensions' of planning was one of the three strands that were deemed to constitute a satisfactory planning education and the importance of equal opportunities was one of the components of the strand. Moreover, within a second, 'knowledge' strand of education, planners

Table 2.2 General preoccupations of planners, 1947–71

Area of interest	Years 1947–51	1952–6	1957–61	1962–6	1967–71
1 Research, related disciplines, outside agencies. Visits, talks, etc.	24.9	13.1	20.1	12.1	11.3
2 Communications with the public/ other groups	1.9	1.5	2.1	1.3	2.3
3 Reform of local authority system	0.0	0.0	0.8	0.0	2.7
4 Regional planning with an inter-regional or economic emphasis	1.9	1.2	1.0	6.0	1.7
5 Interests within the specific area of town and country planning	56.2	64.2	66.0	63.4	56.7
6 Internal reorganisation management	0.5	0.5	0.2	1.6	6.8
7 New techniques concerning information handling and analysis Models, statistics, etc.	0.0	0.5	0.0	1.3	4.0
8 Theories of planning processes, decision-making, etc.	0.0	0.0	0.4	0.5	1.8
9 Planning education, status, recruitment, etc.	0.0	1.7	1.3	6.3	4.0
10 Broad legal and administrative machinery of town and country planning	7.1	3.4	1.8	2.6	2.8
11 Detailed legal aspects and implementation, etc.	2.8	9.5	2.0	1.7	1.6
12 General and miscellaneous, including overseas activities	4.8	4.6	4.4	3.3	4.5

Source: Harrison (1975), p. 267.

Note
This table, based on branch meetings data, shows approximate percentages of time devoted to different general areas of interest within separate five-year periods.

were required to understand 'the distributional implications of public policy' (RTPI, 1991a, pp. 2–3). These are important changes, which must influence the nature of planners' education even where they are honoured in a routinised and lacklustre manner. Moreover, the production of graduates who have been exposed to discussions of social justice and equal opportunities may make some impact on professional practice, although the extent and depth of impact will be qualified by the effects of other influential factors which will be discussed later in this chapter.

The shift in the education guidelines was the product of prolonged debates within the Institute about broadening the concept of planning that it was promoting, from a 'value-free' technical activity focused on plan-making to the management of environmental change, an activity combining technical and evaluative components and which is necessarily conducted in a political context (Hague, 1996). Important as these struggles were (and are), there are serious doubts about the extent to which the Institute's lead in promoting issues such as equal opportunities, in particular, has been followed in everyday planning practice.

Table 2.3 Preoccupations of planners within the specific area of town and country planning, 1947–71

Area of interest	Years 1947–51	1952–6	1957–61	1962–6	1967–71
1 Planning strategies at regional or sub-regional level	2.5	3.9	3.8	16.0	8.6
2 Housing (including rehabilitation and planning of neighbourhoods)	2.5	2.1	4.1	1.1	4.4
3 Urban renewal, redevelopment, central area development	7.6	12.7	13.7	12.4	2.3
4 Structure, local and general Development Plan aspects. Area studies, visits etc.	23.6	35.6	27.9	19.9	18.0
5 National Parks (and Areas of Outstanding Natural Beauty)	1.7	3.2	3.1	0.0	2.4
6 Leisure and recreation planning	6.8	1.6	0.4	3.6	9.7
7 Rural planning and development generally	3.4	2.4	2.8	2.9	8.3
8 Conservation, preservation, reclamation, etc. General amenity planning in rural areas	7.6	7.6	5.9	3.0	17.5
9 Design, civic design, landscape architecture	11.5	9.6	10.2	7.3	2.8
10 Transport, traffic planning	1.7	4.2	11.2	13.4	8.1
11 Explicit social planning (including social service facilities and social problems)	5.1	1.4	0.8	1.6	4.0
12 Retaining, industrial development, offices	0.0	2.7	8.4	1.2	3.3
13 Minerals and natural resources	5.9	3.2	1.9	0.2	1.1
14 Town and country planning matters in general	20.2	9.7	5.7	7.3	9.6

Source: Harrison (1975), p. 268.

Note
This table presents a detailed analysis of interests within category 5 of Table 2.2 (interests within the country planning). It shows approximate percentages of time devoted to different preoccupations for separate five-year periods.

Moreover, the scanty evidence available points to social welfare being a marginal concern in Institute activities at local level (see Tables 2.2 and 2.3).

Harrison (1975) examined the issues discussed by planners at branch meetings between 1947 and 1971. The breakdown presented in Tables 2.2 and 2.3 makes clear how marginal 'social planning' was in the concerns of these planners. No strictly comparable study exists for the period since 1970, but the 1997 attitudes survey of RTPI members suggests that there has been no significant shift in interest towards social welfare issues. When members were asked about what continuing professional development training they would like to undertake, some differences emerged between their preoccupations and those identified by Harrison (see Table 2.4).

There was a striking increase in the amount of interest in planning law and

Table 2.4 Continuing professional development training that respondents would like to undertake

Area of training	% of respondents
Planning law	27.4
Planning for sustainable development	26.5
Management	26.4
Information technology (inc. GIS)	24.9
Environmental appraisal and assessment	20.0
Economic development and urban regeneration	19.3
Urban design	17.1
Conservation (built environment)	16.4
European issues	16.0
Land-use transport issues	14.9
Development control	14.1
Project implementation	13.7
Countryside and rural areas	13.5
Commercial development planning	10.3
Leisure, recreation and tourism	9.8
Environmental protection	9.5
Conservation (natural environment)	8.1
Plan making	7.9
Retail planning	7.5
Enforcement	7.3
Town centre management	6.7
Derelict land	6.0
Minerals and waste management	6.0
Planning and archaeology	5.5
Research	4.6
Coastal planning	3.9
Planning for equal opportunities	3.2
Water and resource management	2.5
Other	2.5
No. of respondents	1,575

Source: RTPI (1997) p. 112.

development control, and considerably less interest in research. However, the greatest single areas of interest would still be captured by Harrison's category of 'specific areas of town and country planning' and, within that, issues of social welfare come well down the list – consider for example, the extremely low level of interest in planning for equal opportunities. This distribution of interest appears to reflect the emphasis and ethos of day to day professional practice, at least, in the local government sector. The 1991 survey referred to earlier asked respondents about the allocation of their time between different planning tasks. Development control, appeals and enforcement took up over 33 per cent of planners' (collective) time, with development planning taking up a further 13.2 per cent. Issues related to access for disabled people (0.8 per cent). women (0.1 per cent) and race relations (0.1 per cent) were insignificant in terms of the proportion of time devoted to them (LGMB/RTPI, 1992, p. 32).

The local constraints on practice will be discussed in a later section of this chapter here; it will be noted that there have been celebrated examples of local planning authorities with strong political direction, which have explicitly addressed issues of distributive justice in their planning programmes (GLC, 1984; Sheffield City Council, 1988), but these appear to be exceptional. Healey's exhaustive survey of local plans concluded:

> Whatever their stated purposes, most local plans contain proposals primarily about land allocation. This partly reflects central government advice that proposals should be limited to those which can be achieved by planning powers. Yet this limitation applies to non-statutory plans as well.
>
> (Healey, 1983, p. 189)

There is no reason to suppose that matters have changed since. For example, first-hand accounts report either indifference to, or a blithe complacency about, the social good that planning achieves (Nicholson, 1991). Yet this is a conservative complacency, for a number of surveys have reported planning departments finding it difficult to sensitise working practices and policies to the principles of promoting equal opportunities (L. Davies, 1996; Krishnarayan and Thomas, 1993; Little, 1994a).

Perhaps these findings are not surprising given the importance of judicial interpretations of planning law and central government policy in shaping professional practice (McAuslan, 1980; Tewdwr-Jones, 1995). Law and policy are important because they define what are proper concerns of the planning system – i.e. what constitutes a good or proper reason, in planning terms, for taking action. Put crudely (and a little misleadingly) they define what planning professionals 'can or cannot do'. Both law and policy have changed from time to time over the last fifty years and at any one time there is always a degree of uncertainty about the precise boundaries of what constitute planning reasons (Cullingworth and Nadin, 1997; Healey, 1996a; McAuslan, 1980, ch. 6). Yet whatever the uncertainties at the margins there is no doubt that the central thrust of government policy and judicial interpretation of planning law (the former sometimes following the latter, see McAuslan, 1980, pp. 173–4) has been to promote a view of planning as primarily concerned 'with physical matters closely connected to the land and its development' (McAuslan, 1980, p. 155), with no (or scant) consideration of who, or which groups, benefit from development. Occasionally, such considerations have forced their way into planning, notably in relation to securing a supply of low-cost housing in areas of development pressure and/or restricted supply. The difficulties such issues provoke from practising planners is perhaps in part an indication of how unusual it is for distributive questions to be addressed in the planning system.

Explanations of marginalisation

How are we to explain the marginalisation of social welfare within the concerns of the planning profession? An explanation will be offered which focuses on the material circumstances within which planners have undertaken their work. Such an approach can be contrasted with those which emphasise the significance of competing ideologies (McAuslan, 1980), or which stress the origins of the profession and/or the academic background of current entrants. It will be argued that three factors are especially important to the explanation offered here.

The first factor is the strategy for professionalisation followed by the planning profession. Professionalisation involves an attempt on the part of an occupational group to get as close as possible to monopolising the provision of particular services, in order to provide itself with leverage in setting the terms on which the services will be provided. The most popular and successful way to gain such a position in twentieth-century Western countries has been to set up a 'qualifying association' (i.e. a professional institute), membership of which is necessary (either *de jure* or *de facto*) or very useful for gaining employment in a given field. The Royal Town Planning Institute is such an association, and its history contains a series of discussions, continued to the present day, about how it should define its area of expertise and regulate entry to its membership (Cherry, 1974; Faludi, 1972; Hague, 1984, 1996). The Institute originated as an offshoot of three professions which presented themselves as having a clearly defined technical expertise: architecture, surveying and engineering (Cherry, 1974). Such origins might well have influenced how the aspiring profession sought to present itself in its early years, but cannot plausibly be held to have determined the trajectory of any strategy in the longer term. Nevertheless, the Institute *has* tended to emphasise a technical, value-free conception of planning for most of its existence, for such a view has great practical advantages for an occupational group seeking to monopolise the provision of a service. As Larson (1977) points out, the claim to knowledge-based expertise is a powerful argument for monopolising a service and such a claim is more authoritative in technical activities than in those that involve moral political or aesthetic evaluations. For example, few doubt the specialised abilities of architects to create buildings which function well, but their claims to exercise aesthetic judgements on behalf of others are challenged regularly. For the planning profession, too, it has been, and remains, useful to present its expertise as technical and quasi-scientific and consequently not dwell on such questions as what constitutes a good or just society and how town planning can help to bring it about. The most bitter criticisms of planning have arisen from occasions when it has become (often wrongly) identified with (failed) attempts to create better lives for people, such as high-rise housing. It is significant, for example, that Chadwick commended the systems view of planning to the members of the RTPI on the grounds that it provided a technical 'rationale' for professionalism:

> planning is concerned with very large . . . systems . . . It is the planner's concern to recognise and understand the system as a whole and to predict its future optional performance. Here is the justification for planning as the separate professional activity that undoubtedly it is.
>
> (Chadwick, 1966, p. 186)

He contrasted this with (then-current) 'puerile enthusiasms' for the content of plans (i.e. creating particular kinds of places):

> If a focus on the rationality and technicality of planning has helped in sustaining the profession's credibility to a general public, it has been even more significant in bolstering the profession within local government.
>
> (Ibid., p. 184)

Healey and Underwood (1978, p. 106) found that London planners felt that the most important influences on the scope and nature of local planning were the councillors and internal administration – i.e. party politics and organisational politics at the local level. More recent case studies confirm this impression (Kitchen, 1997) and highlight the pressure (and support) which local politics can exert on planners (Lees, 1993; Krishnarayan and Thomas, 1993). It is important, therefore, to examine the local government context of plans. As stated earlier, the planning profession is largely a local government profession that has grown as part of the enormous development of the public sector at central and local levels in Britain post-1945. The growth of local government has been prompted and shaped by a welter of legislation placing duties on local councils; specialised occupations have grown up to undertake these duties and many (perhaps most) have sought to improve their status and prospects by professionalising (i.e. controlling access to a certain range of jobs). Town planners have been one of the more successful groups engaged in this process, but it has been (and remains) necessary to defend the boundaries already established and, if possible, extend them to include new local government activities (e.g. planners are one of the occupational groups active in the new environmental obligations placed upon local authorities). Healey and Underwood's research shows that in the bruising world of organisational politics planners can rarely sustain claims to roles which involve coordination of, or control over, the activities of other professions. In recent decades they have had some success in colonising new activities which fall between traditional departmental responsibilities in local government – e.g. in relation to tourism promotion, economic development and aspects of urban policy (Thomas, 1984) – but often these forays are temporary incursions before departmentalism engulfs the new policy area. Consequently, the most consistently successful strategy has been continually to bolster and, if necessary, fall back upon a role definition which has emphasised core activities sanctioned by legislation (development planning and development control) and, more generally, the control and management of land use. These have the virtues of not encroaching on the domains of other professional groups in local government (though 'border skirmishes' may occur from time to time) and of lending themselves to being presented as technical skills and aptitudes (even if the skill sometimes consists of simply knowing how the planning system works – see Thomas, 1992).

The final factor which has tended to push social welfare to the margins of planners' concerns has been planning law and, more particularly, its judicial interpretation. The law is at the heart of the key planning function of development control and in as much as it defines what constitute valid planning reasons for action it is widely considered by planners to define the 'essence' of planning. As mentioned above, though there are often disputes about legal interpretations, there can be little doubt that neither legislators nor judges have considered the mainstream of planning law to be the vehicle for addressing social and economic inequalities. The New Town programme, it is true, spawned plans that, in a variety of ways – in different towns – sought to create what were considered to be better ways of life (e.g. through social mix, reduced dependence on private cars and employment opportunities in 'modern' industries). But influential as New Towns might be as shop windows for British planning, they remained a sideshow in terms of the numbers of planners they employed and the working practices they employed (Ward, 1994). New Town corporations were not organised in the same ways as local authorities, nor were they subject to the same constraints or premises (of course, they had their own to contend with) (Pell, 1991). This meant that they would never be central to the self-perception of professional planners in Britain.

These remarks on the legal framework that defines and sustains state intervention in the land market and the process of development lead us to consider the interpretation of the law. If judicial interpretation of planning law were ever seriously inimical or threatening to government intentions, then the law could be changed or clarified so as to ensure that policy intentions were fulfilled. There are some examples of this happening, such as the insertion (in 1986) of S19A into the Race Relations Act of 1976, in order to bring planning explicitly within its remit, after a judicial decision which suggested that public activities such as planning might not be covered by the Act as it stood. But tension and feuding between governments and the courts is not a feature of the history of planning (Ward, 1994); indeed, as McAuslan (1980) points out, it is not unusual for government to defer to judicial interpretation of the law and to frame policy advice to local authorities accordingly.

This conservatism in defining the role and scope of planning is only surprising if we believe that planning, as an activity of the state, is bound up with the promotion of a more just and egalitarian society. Rees and Lambert (1985) have noted the influence of what they term the Whig view that modern planning's emergence can be explained as part of a continual and inevitable story of social progress, to which the practice of planning itself contributes. A more plausible mode of analysis, of both the origins and contemporary practice of planning, grounds it in the social and economic tensions associated with modern capitalism. Accounts which follow this approach see no inevitability about planning being guided by any particular principles, or seeking particular ends. Rather, the scope, content and direction of planning are shaped by political struggles, at various spatial scales, in which the protagonists (and lines of cleavage) arise from the conflicts of interest endemic in capitalist society. The nature of planning will be shaped by those interests which wield consistent and widespread influence within British society. Some have argued, persuasively, that this has meant that planning has never systematically threatened prevailing social and economic inequalities, but has, in general, served the purpose of *managing* land markets and the development process, for example, by dampening the volatility of markets through curbing excessive speculation (Fogelsong, 1996; Griffiths, 1986; Harvey, 1996). This kind of account of the role of planning in modern Britain would not find it paradoxical that social welfare is so marginal to the planning profession. On the contrary, its peripheral status is to be expected in an occupation engaged in an essentially conservative function.

3 Town planning

'Social', not just 'physical'?

Nigel Taylor

> *Town and Country Planning might be described as the art and science of ordering the use of land and the character and siting of buildings and communicative routes. . . . Planning, in the sense with which we are concerned with it, deals primarily with land and is not economic, social or political planning, though it may greatly assist in the realisation of the aims of these other kinds of planning.*
>
> (Keeble, 1952, p. 1)

Physical and social: town planning theory since 1945

'Planning' in the general sense of the state intervening in and managing social and economic affairs (i.e. not just 'town' planning) attracted widespread support in Britain towards the end of the 1930s and during the Second World War. The Barlow Report (Barlow Commission, 1940) was a significant document in this respect; it was driven as much as anything by economic considerations and in particular, by a concern for the economic inequalities between the regions of Britain which became apparent in the economic depression of the inter-war years, as manifest particularly in the disparities in unemployment. The 1945 Distribution of Industry Act was the outcome of Barlow and is often recorded in histories of town planning as one of the quintet of post-war planning Acts.

Certainly, town planning has long been seen as having a regional dimension and this was the case in the immediate post-war years. Thus Keeble's planning textbook *Principles and Practice of Town and Country Planning* (1952) covers regional planning as well as the planning of towns within regions. However, responsibility for administering the regional controls over industrial location that were introduced by the 1945 Act came under the control of the central government Board of Trade, not the (then) Ministry of Town and Country Planning, and this is significant. It indicated how 'mainstream' town and country planning was not viewed as an activity responsible for strategic economic planning and management, but rather as one concerned with the more detailed planning of urban development. In particular, town planning in the strict sense was viewed as being primarily concerned with planning the physical shape and development of towns and cities.

This view of the primary focus of town and country planning was exhibited in

Keeble's book and indeed in most other texts on town and country planning of the early post-war period. Thus, in spite of his references to regional planning, Keeble took the view that town planning was essentially an exercise in *physical* planning, as the quotation at the head of this chapter makes clear. I have described this post-war 'physicalist' conception of town planning in more detail elsewhere (Taylor, 1998, ch. 1), but here it is worth drawing attention to two things which accompanied it. First, the view that town planning was about the physical planning of the environment was intimately connected with the view that it was essentially an exercise in (physical) *design*. Indeed, town planning at this time was seen (as in fact it had been seen throughout human history) as a natural extension of architecture or (to a lesser extent) civil engineering. Accordingly, those thought to be most appropriately qualified to be town planners were architects or civil engineers, or at least people qualified in the built environment professions, such as surveyors (Cherry, 1974, p. 210). Second, the view of town planning as an exercise in physical design led to a certain view of the kinds of plans it was assumed appropriate for town planners to make, namely, they should make 'master plans' for towns which delineated precisely the future pattern of land use and development for the urban area being planned. In other words, town plans were conceived as *blueprint* plans for future urban land use and form, on the same model as architects' plans for buildings or civil engineers' detailed designs for roads or bridges (cf. Faludi, 1973, ch. 7) In sum, the view of town planning advanced by Lewis Keeble and most other post-war planning theorists was one which emphasised the physical shape and form of urban development.

By the early 1960s this physicalist theory of town planning was being criticised in the light of the experience of the first fifteen years of post-war planning practice. From these criticisms there emerged new theories of planning which sought to embody a conception of planning that was more socially informed and which, correspondingly, down-played the physical design side of town planning. The slogan 'planning is for people' emerged at this time. The expression was so over-used that it soon became a meaningless cliché, but it did at least express the shift from a physical, design-based concept of town planning to one that was more socially informed and sensitive. In this section, I shall recount three episodes in post-war planning thought which led to the development of more socially informed town planning theory. These are: first, the lessons learnt from the social insensitivity of post-war physical planning, especially the schemes of comprehensive housing redevelopment; second, the emergence of local urban protests against planning schemes in the 1960s and the consequent acknowledgment of the political nature of town planning; and third, the realisation, in the 1970s, that town planning practice affected different social groups in different ways and so had 'distributive' effects that could diminish or heighten social inequalities (c.f. Dennis, 1970, 1972).

The social insensitivity of post-war physical planning

Britain entered the post-war era with a huge legacy of old Victorian terraced housing in the inner areas of its major industrial cities. The appalling condition of much of this housing led whole areas to be declared 'unfit for human habitation' and, in consequence, designated as 'comprehensive development areas'. The 1950s therefore saw the beginnings of massive inner-city 'slum clearance' and comprehensive redevelopment schemes. Whatever the justification for such schemes in terms of physical, environmental conditions, the process of comprehensive redevelopment and rehousing inevitably resulted in huge upheaval for the inhabitants of these areas. Whole communities were uprooted, and by the end of the 1950s this 'clean sweep' physical planning was being criticised for its social insensitivity.

The very physicalist theory of planning that held sway at this time was itself a major cause of the social insensitivity of early post-war planning, for in concentrating on the physical and design aspects of the built environment, planners were blind to the nature or quality of the social life that was lived in the areas they were replanning. Thus in their famous study of housing redevelopment in a part of east London, the sociologists Peter Young and Michael Willmott (1957) revealed how, in focusing on the inadequate physical conditions of the nineteenth-century 'slum' housing and the aim of rehousing people in physically better accommodation in physically better environments, planners and other officials seemed not even to notice the social communities that thrived in these areas. The proposals for comprehensive redevelopment and rehousing were therefore based on physical, but not social facts about these areas (Dennis, 1970).

However, it would not be true to say that post-war physical planning theory and practice paid no regard at all to questions about social communities. On the contrary, although town planning theorists and practitioners paid scant regard to real-life social communities, they did see themselves as engaged in helping to foster social communities through the activity of physical planning. In particular, in planning residential areas, planners aimed to create geographically distinct 'neighbourhoods' planned in such a way as to nurture the formation of local social communities. Thus, in the master plans for all the post-war New Towns, or in Patrick Abercrombie's plan for London, the town was given a neat cellular structure of residential neighbourhoods, each of which was designed to be relatively self-contained and socially balanced. By providing each neighbourhood with 'its own' local facilities, such as local shops, a local primary school, a local park, a church and community centre, and by siting these facilities at the geographical centre of the neighbourhood, planners hoped to facilitate opportunities for the inhabitants of neighbourhoods to meet and, in time, form a social community.

In spite of this overt concern for 'community', neighbourhood planning ideas exposed the physicalist bias of town planning thought and practice in another way, for in believing that social communities could be nurtured by physical means, town planners once again exhibited their ignorance of real-life social communities. As sociologists like Maurice Broady pointed out, it was mostly *social* conditions such as long-term residence in an area or the experience of economic hardship that explained the existence of local social communities, rather than physical environmental conditions (Broady, 1968, pp. 14–15). In seeking to create social communities by physical means, therefore, town planners were relying on a naive and essentially false theory of physical or architectural 'determinism' – so in concentrating on physical environmental conditions, post-war town planning displayed either an insensitivity towards, or an ignorance of, urban social conditions.

In fact, town planning theory and practice at this time exhibited a general lack of understanding of the social and economic life of cities and it was this that was at the heart of Jane Jacobs's powerful attack on town planning orthodoxy at the beginning of the 1960s. As she put it:

> Cities are an immense laboratory of trial and error, failure and success. . . . This is the laboratory in which city planning should have been learning and forming and testing its theories. Instead the practitioners and teachers of this discipline . . . have ignored the study of success and failure in real-life, have been incurious about the reasons for unexpected success and are guided instead by principles derived from the behaviour and appearance of . . . suburbs, tuberculosis sanitoria, fairs and imaginary dream cities – from anything but cities themselves.
>
> (Jacobs, 1964, p. 16)

Essentially the same point was made by Christopher Alexander (1965) in his celebrated article 'A city is not a tree'. According to Alexander, real cities possess complex structures containing numerous relationships and 'overlaps'. But the planning of cities into, for example, self-contained neighbourhoods, or single-use districts, exhibited an over-simplified model of the city. Here again, then, town planning was criticised for lacking an understanding of the socio-economic richness and vitality of existing cities.

It was out of these criticisms of post-war physical planning theory that there developed, in the 1960s, alternative theories of planning which sought to take account of the socio-economic functioning, not just the physical structure and form, of cities. The systems view of planning was in part an expression of this shift in planning thought. The starting point of the systems view was the obvious one that planners should understand the object they were planning before they engaged in planning it (cf. Geddes, 1915). The systems theorists then advanced the view that this object – cities, regions and the environment generally – was best seen as a functioning system of interconnected activities. As Brian McLoughlin, the leading British systems theorist, put it: 'The components of the [urban/environmental] system are land uses and locations which interact through and with the communications networks' (McLoughlin, 1965, p. 260).

Systems theory had in fact been borrowed from disciplines other than town planning, notably the highly technical areas of operations research and cybernetics. As a result, when systems theory was applied to town planning it came with the technical baggage of mathematical modelling and techniques of system optimisation. As such, it ended up looking far removed from the ideas of writers like Alexander, Jacobs and Broady. Yet the essential idea of the systems view of planning was in accord with what these critics of post-war planning theory had been saying, namely that the reality of cities was much more complex than planners had hitherto assumed and that planners therefore needed to develop their understanding of this complex reality before blundering in with plans to change it. And with its stress on understanding cities (or regions, etc.) as systems of interconnected activities, the protagonists of the systems view of planning advanced essentially the same view of cities as Christopher Alexander. More to the point in the context of this discussion, the systems view implied that town planners needed to understand the *social and economic* functioning of cities, rather than seeing cities just in terms of their physical form and appearance.

The emergence of urban protest and the politicisation of town planning

By the end of the 1960s, the kind of social insensitivity displayed in the 'clean sweep' approach to town planning, most notably in the schemes for comprehensive housing redevelopment and new urban motorways, had provoked some communities into open rebellion (see J. G. Davies 1972; Goodman, 1972). The 1960s therefore saw the emergence of local urban protest movements and from this a very important theoretical lesson was learned: namely, that town planning judgements were not just technical, but rather judgements of value about the sort of urban environment it was desirable to create. In this respect, the technicism of some versions of the systems view of planning was as open to criticism as that of the design-based conception of town planning that preceded it. Further, since planning judgements affected the lives and interests of whole communities, they were also clearly *political* value-judgements.

As a result of these urban protest movements, there developed through the 1960s a concern for, and in Britain statutory provision for, public participation in town planning. This represented an open acknowledgment that town planning is a value-laden political activity. There was, of course, debate about how far the relevant public should participate in planning decisions (DoE, 1969; Arnstein, 1969) and hence what the very notion of 'public participation' meant or implied in terms of citizen power over planning decisions. There was also debate about what the role of the town planner should be in relation to the public. Thus Paul Davidoff (1965) argued that the planner should see himself more as an advocate for the views of different groups in the public, rather than as a technical expert with a superior view of 'what was best' (cf. Macpherson, 1977). This is not the place to discuss these debates further. The main point to note in the context of this chapter is that, by the end of the 1960s, town planning was being described as a political activity and this represented a further aspect of the development of a more socially informed concept of town planning.

The distributive effects of planning

Planning theorists like Jane Jacobs and Christopher Alexander had called upon planners to gain a better understanding of cities before they embarked upon planning them. In the same vein, a number of other theorists called for the empirical examination of planning practice itself (e.g. Scott and Roweis, 1977), so that we might better understand its effects and its role in society. Yet up until the 1970s there were few, if any, systematic empirical studies of the effects of town planning in practice. Against this background, the publication in 1973 of the two volume work *The Containment of Urban England* by Peter Hall and his colleagues was a significant event (Hall *et al.*, 1973), for Hall and his fellow researchers had set themselves the task of examining the changing face of urban Britain during the twentieth century and, as part of this, the operation and the effects of the post-war British planning system down to the end of the 1960s.

Hall and his colleagues observed that a major objective of the post-war planning system had been to restrict urban sprawl and development in the countryside and they found that this objective had been realised: hence the title of their work. However, they believed that their research showed that urban containment had been bought at a price. By restricting the supply of land available for urban development, British town planning had unwittingly contributed to the inflation of land and property prices. This, in turn, had led town planning to have some distinct distributive effects on different groups within British society. In particular, by contributing to increases in land and property prices, physical planning had contributed to the widening of social inequalities. Those people who were materially better off, and who therefore owned or could afford to buy property, benefited from property price inflation, whilst those who were materially worse off obviously found it more difficult to buy their own homes or had to put up with inferior quality council housing. In short, post-war physical planning had had socially inegalitarian, or socially 'regressive', effects.

The conclusions of *The Containment of Urban England* could be disputed. They implied that town planning practice was a significant agent in shaping urban development and hence in causing the aforementioned socially distributive effects. This was questioned by a number of urban theorists who argued that the actions of town planning had to be seen within the political and economic context of market capitalism and that when public sector planning was seen in this light, it was not such a significant determining agent as Hall and his colleagues had assumed (see, e.g. Rees and Lambert, 1985; cf. Fainstein and Fainstein, 1979, p. 148). As Pickvance (1977, p. 269) contended: 'the determining factor in urban development is the operation of market forces subject to very little constraint'. Once again, it is not my concern here to explore this debate further. Rather, my concern is to show how, in the 1970s, a recognition that town planning could have (more or less significant) distributive effects contributed another dimension to the development of a more socially informed town planning theory.

So much, then, for this résumé of (some) developments in town planning theory since 1945. The picture which emerges is that, whereas in the years immediately following the Second World War town planning was conceived of as essentially an exercise in physical planning and urban design, through the 1960s and 70s – largely as a result of criticisms of the physicalist model in practice – various theorists drew attention to the need for town planning to become much more 'socially informed'. But does it follow from this that we should now speak of town planning as being 'social' as well as (or even rather than) 'physical' planning? Have the developments described in this section rendered the old talk of 'physical' planning and relatedly of town planning as 'design', completely redundant? What, if anything, remains of Keeble's notion of town planning as 'physical' planning? And, assuming town planning has *something* to do with the physical environment, how should we conceive of the relation between the physical and the social in town planning? It is to these questions that I now turn.

Town planning: 'social', not just 'physical'?

In this section I adopt a mode of analysis quite different from the descriptive, historical overview of the previous section. With that history as background, I turn now to a more reflective consideration of what might be meant when it is suggested that town planning is 'physical' and/or 'social' and how physical and social planning are related. Put another way, what I offer here is an analysis of the concepts 'physical' and 'social' as they are (or might be) employed as descriptors of town planning, together with the logical relations between these concepts. Readers unacquainted with (or perhaps unsympathetic to) such conceptual analysis might dismiss this exercise as 'merely semantic' and/or 'merely formal'. To this I make two replies. First, what is offered here is, unashamedly, a piece of semantic and formal logical analysis; it is an analysis of the *meanings* of the terms 'physical' and 'social' in relation to town planning and of the logical relations between them. Second, this analysis is not 'merely' semantic or formal. For words (or concepts) and their meanings refer to actual things and in clarifying the foregoing terms and their logical connections we are simultaneously clarifying what kind of practice town planning is.

To ground this exercise, I shall take as the basis for my analysis the definition of town planning advanced by Lewis Keeble which was quoted at the head of this chapter. I paraphrase it here again as follows:

> Town (and country) planning is the art and science of ordering the use of land and the character and siting of buildings and communicative routes. Town planning deals primarily with land and, as such, it is not economic, social or political planning. But town planning can assist in realising the aims of these other kinds of planning.

From this we can distil the following propositions or claims:

K1 Town planning deals primarily with land.
   (K1 is the major premise)

K2 Town planning is concerned with planning/ordering:

(a) land use and
(b) the character of buildings and spaces (including communicative routes).
(K2 is an elaboration of the major premise K1)

K3 Town planning is not economic, social or political planning.
(K3 appears as a logical consequence of K1 and K2 together)

K4 Physical town planning can contribute to the achievement of economic, social and political goals.
(K4 is a separate empirical claim)

As noted here, the major premise in Keeble's statement is that town planning deals with land (i.e. K1 above). K2 is really an elaboration of this, in two parts. One part of this elaboration is that town planning is concerned with planning (ordering) the *character and siting of buildings and communicative routes* (K2b above) and I suggest that it is in respect of this proposition that, in Keeble's view, town planning is primarily an exercise in *physical planning and design*. Further, it is from this proposition that Keeble concludes that town planning is not economic, social or political (K3 above). However, Keeble makes the separate, empirical claim that physical planning can contribute to the realisation of the aims of these other kinds of planning – that, in other words, physical planning can realise social (as well as economic and political) ends (K4 above).

Given that K2 is an elaboration of K1, I suggest that we can further reduce Keeble's statement on the nature of town planning to the following three propositions or claims:

C1 Town planning is primarily physical planning.
(C1 is a major definitional claim)

C2 Given C1, town planning is not social planning.
(This is a logical inference from C1).

C3 Physical planning has social effects and so can realise social goals.
(This is a separate empirical claim)

In this reduction, I have confined myself to 'social' (as distinct from 'economic', or 'political') effects and goals because this is my concern in this discussion (although I shall later suggest that we cannot strictly distinguish between 'social', 'economic' and 'political' matters in the way Keeble's statement implies). I now turn to an examination of claims C1–C3.

C1: 'Town planning is primarily physical planning'

This is a claim which Keeble makes 'by definition'. However, the statement is ambiguous because it is open to two possible interpretations. Thus it is unclear whether the statement is to be taken as an empirical or a normative claim, in other words, whether C1 is a definition of what town planning actually *is* (i.e. in practice), or whether this statement represents Keeble's considered view as to what town planning *should* be like. We therefore have two possibilities:

C1.1 'Town planning *is* (as a matter of fact) primarily physical planning'
C1.2 'Town planning *ought to be* primarily concerned with physical planning'

Let us consider these in turn.

C1.1: 'Town planning is primarily physical planning'

This is an empirical claim (a claim about what is the case, as a matter of fact). However, it still remains unclear whether it is a claim about planning *as an activity*, or about the things that town planning deals with, the *object* of town planning. Call the former C1.1a and the latter C1.1b.

Consider, first, C1.1a. This is a claim about the kind of activity that town planning is (as a matter of fact). But straightaway it would seem odd to describe the *activity* of town planning as 'physical', even though the activity involves handling physical objects (maps, plans, reports, etc). For the main thing about town planning 'as an activity' is that it is carried out by human beings; indeed, we only attribute human agency to human beings and so it is *only* human beings who engage in activities like town planning. This being the case, it seems better (i.e. more accurate) to describe the *activity* of town planning as a form of social action. But if it is a form of a social action, it is not thereby just 'physical'. Claim C1.1a is therefore false.

Consider, next, C1.1b. This is an empirical claim about the object of town planning and it says that that object is – primarily – physical. For example – to go back to Keeble's original statement – it is about the 'character and siting of buildings and communicative routes' and these are physical objects. Now, herein lies the germ of important truth in Keeble's whole statement. For, as actually constituted when Keeble wrote his book (1952), town and country planning in Britain had powers primarily over the location and form of physical development, and this has remained the case down to the present day. Thus in spite of the numerous changes to planning legislation since the Town and Country Planning Act of 1947, it is still true that the object of town planning – the thing that it seeks to manipulate – is physical development. C1.1b is therefore largely true. I say 'largely' true deliberately, for it is not wholly true. Town planning in Britain, as established in 1947 and still now, has powers to control (certain) changes in land use, for all the planning acts since 1947 have defined 'development' not just in terms of changes to physical form, but also in terms of changes in land use. To be sure, most changes in land use issue in changes in physical form of some sort, but not all do. For example, I might change my house to an office block without any change in its external physical form or appearance. Likewise if I converted my house to a pub (apart from a sign). Yet, ever since 1947, in Britain both these changes have constituted 'development'. So, if we are being precise, claim C1.1b is not strictly true either. For in planning land use, British town planning since 1947 has been concerned with planning human activities, not just 'the character and siting of buildings and communicative routes' – not just physical form. However, remember that C1.1b says that the object of town planning is *primarily* physical – that town planning is, with regard to its object, *primarily* physical planning. With this qualification in place we can accept C1.1b as true.

C1.2: 'Town planning ought to be primarily concerned with physical planning'

This is a normative claim. Therefore, to the extent that there is any dispute over what the fundamental purposes of town planning should be, this claim must be open to dispute – and indeed, it has been disputed. In 1960, Donald Foley drew attention to the fact that there were different views about what town planning should be trying to achieve and further, there were tensions between these different views (i.e. they were not necessarily compatible). Foley described three contrasting views or 'ideologies' of town planning. The details of his account need not detain us here, but the second and third of Foley's ideologies touched on the issues relevant to the above claim. Thus Foley's second ideology suggests that 'Town planning's central function is to provide a good (or better) *physical* environment' (p. 77 in Faludi, 1973; my italics), whereas his third ideology sees town planning 'as part of a broader *social* programme' (ibid., p. 78; my italics).

In fact, it is worth noting that, in Foley's account, the third view or ideology does still seem to assume that the object of town planning is the 'physical' environment. It is suggested that town planning is part of a broader social programme only in so far as 'it is responsible for providing the *physical* basis for better urban community life' (ibid., p. 78; my italics). Be this as it may, there is no reason in principle why someone should not maintain that town planning ought to have a wider remit than just a concern with land use and physical development – that it should also be concerned, directly, with planning the economic, social or political life of a town. For example, town planning could conceivably encompass the setting up of a particular system of political decision-making for a town (say, a participatory one), or a system of communal landownership. Indeed, Ebenezer Howard's proposals for garden cities a hundred years ago seem to have envisaged this wider concept of town planning.

Whether or not town planning should have this wider remit, or be confined ('primarily') to physical planning, is a matter of ideological dispute, as Foley's use of this term makes clear. For example, some socialists might advocate a form of town planning that encompassed the political and economic matters instanced in the previous paragraph. Free market liberals, on the other hand, would probably resist such an extension of the scope of town planning. Indeed, some liberals have opposed the institution of public sector town planning, including the planning of physical development, altogether (e.g. Jones, 1982). It is not my purpose here to explore further such ideological debate, still less to broadcast my own view of what the scope of town planning should be. Suffice it to note that:

1   C1.2 is a normative proposition concerning the proper remit or scope of town planning;
2   as such, C1.2 raises fundamental questions about not only the remit of town planning, but also about whether there should be any system of public sector town planning at all;
3   one's view about the scope of town planning depends on one's moral and political ideology.

C2: *'Town planning is not social planning'*

To simplify the analysis, let us henceforth operate with claim C1.1b discussed above. In other words, let us take claim C1 to be an *empirical* claim about the object of town planning. C1 therefore says that town planning is (as regards its object) primarily physical planning. Our discussion has shown this claim to be true of British town planning since 1947 (subject to the qualification that some changes in land use that are controlled by planning are not always correlated with significant changes in physical form). Now, given that this is what Keeble meant by C1, Keeble advances C2 as if it were a logical inference from C1. Thus, from the proposition that town planning is primarily concerned with physical planning, it is inferred that town planning is not social (or economic or political) planning. However, as with C1, this statement is open to both empirical and normative interpretations. Thus C2 might mean:

C2.1:   'Town planning *is not* social planning' (as a matter of fact, e.g. in Britain)

Or C2 might mean:

C2.2:   'Town planning *ought not to be* social planning'

Once again, I shall take each of these possibilities in turn.

C2.1: *'Town planning is not social planning'*

The first thing to be clear about here is the logical relation of this proposition to proposition C1 in its empirical form (i.e. C1.1b). At first blush, the logic seems to be as follows: if it is empirically true that (in Britain, say) town planning is primarily physical planning, then town planning cannot be social planning, for town planning's being primarily 'physical' would seem to rule out its being anything else, including 'social'. However, this only follows if what is physical cannot at the same time be social – if, that is, the categories 'physical' and 'social' as applied to defining town planning are mutually exclusive. But they are not necessarily mutually exclusive. Whether or not they are so depends on what we mean by 'social' in this context, and my discussion of this takes us to the heart of this chapter.

There are three relevant senses in which the term 'social' might be used as a descriptor in relation to town planning. First, the term 'social' might be used to describe the *kind of action* that town planning entails: namely, it is a form of social action. In this sense, then, C2.1 would mean: 'Town planning is not a form of social action'. Call this C2.1a. Second, the term 'social' might be used to describe the *object* of town planning; that is, what town planning deals with. In this case C2.1 would read: 'The object of town planning is not social'. Call this C2.1b. Third, the term 'social' might refer to the *purposes* of town planning. On this last reading, C2.1 would be saying: 'The purposes of town planning are not social'. Call this C2.1c. I shall consider each of these three possibilities in turn.

C2.1a, then, suggests that town planning is not a form of social action. Plainly, this is false, certainly as applied to institutionalised statutory town planning in Britain. For action by the state is necessarily collective action, and collective action is social action. In fact, this point was made before when I discussed C1.1a, so what was said there applies here too.

What about C2.1b? This claims that it is the *object* of town planning that is not social, i.e. that the thing(s) that it deals with and manipulates are not social. Arguments about this could become complex, for there are senses in which it could be argued that town planning does, so to speak, act on people. But town planning does not do so in the same direct way as, say, teaching or medical care act on people. Both education and healthcare deal directly with people: people, their minds and bodies, are the objects of education and healthcare. By contrast, as I have said before, town planning works (primarily) on the physical environment. Hence the truth of C1.1b discussed earlier. It is here, then, that town planning's being physical makes it not social. Summing up claims C1.1b and C2.1b together, we can conclude that it is empirically true that the object of town planning is primarily physical, and also that its object is not social.

But does it follow from this that the *purposes* of town planning are not social? This brings us to C2.1c. We can get a firm grip on this question if we ask ourselves why we engage in town planning, or, more bluntly, what is the *point* of town planning? And the answer, surely, must be that we assume or believe that town planning in some way improves the quality of people's lives by improving (or maintaining) the quality of the physical environments they live in. We don't plan the physical environment for its own sake. The *purpose* of planning the physical environment is therefore human or social, not physical. This is true, for example, of planning or controlling the design of buildings and spaces, which might be thought by some to be a part of town planning that is 'not social'. Certainly, some who have written about town planning sometimes talk of town planning having 'social and economic' aims *as well as* aims concerned with physical form and design, as if these latter were not 'social' (or 'economic'). But again, the reason we are concerned with the design quality of the built environment, and hence the *point* of planning it, is that it enhances human welfare, and this is a human, social reason. Indeed, the idea that we might engage in physical planning for non-social reasons is incoherent. So, if the purposes of town planning are social, claim C2.1c is false.

I turn, now, to claim C2 in its normative form.

C2.2: 'Town planning ought not to be social planning'

Once again, the claim that town planning *ought not to be* (or – for that matter – *ought* to be) social is open to different interpretations, depending on what sense we ascribe to the term 'social' here. And once again, the term could be interpreted in the three ways described in the previous section, namely, as a description of town planning as a *form of action*, a description of the *object* of town planning, or a description of the *purposes* of town planning. As in its empirical form (C2.1), therefore, we need to consider claim C2.2 in these three different guises. Call these three different normative versions of C2.2, C2.2a, C2.2b and C2.2c (these correspond to the three empirical forms of C2 considered in the previous section: C2.1a, C2.1b and C2.1c).

Take C2.2a, the claim that town planning ought not to be social planning. At one level, this is a non-starter, because as an institutionalised activity of the state town planning necessarily is social. Given this, to claim that town planning ought not to be social would be incoherent. However, matters are not quite so simple. For someone might argue that it would be better if town planning were not an institutionalised activity carried out by the state, that we ought not to have this form of social action and that, instead, environmental change and development should be governed entirely by private (or 'privatised') individual decision-making. Such a position might be argued for by, for example, extreme classical liberals, or maybe some individualistic anarchists. So at this level, claim C2.2a becomes a matter of ideological dispute about the proper role of the state – or social action – in relation to the fashioning of the built and natural environment. Here, therefore, we return full circle to our earlier discussion of claim C2.1, which held that town planning ought to be primarily concerned only with physical planning. Debate about the efficacy of C2.2a is the mirror image of that same ideological debate.

Consider, next, C2.2b, the claim that the *object* of town planning should not be social. Earlier, we established that, as presently constituted in the United Kingdom, it is *empirically* true that the object of town planning is primarily physical, and not social (see the discussion of C2.1b above). However, it is possible to envisage a form of town planning that would deal directly with social as well as physical objects (a nasty example of this was Ceausescu's Rumania). This, again, is a question about what the scope of town planning should be – what it should encompass – and again, this is a matter of ideological debate, about which adherents to different political ideological positions would differ.

What, finally, of C2.2c, the normative claim that the *purposes* of town planning should not be social? We established earlier (when discussing C2.1c) that the purposes of town planning necessarily are social, that town planning is only engaged in for social purposes, and not 'for its own sake'. From this it follows that the claim that the purposes of town planning should not be social is incoherent, for those purposes necessarily are social. Furthermore, it is worth adding that this is true even if one held that town planning should not be a form of social action, that there should be no institutionalised state-run form of town planning, as some extreme liberals or anarchists might propose. For even if town planning were not a social activity, but carried out entirely by private individuals in a free market, the reasons liberals or anarchists would favour this individualised form of town planning would likely be social, i.e. they would argue that this form of town planning improved social welfare more than town planning conceived and established as a form of social action.

So much, then, for the various forms of the claim that town planning is not social. We come, finally, to the third of the claims advanced by Lewis Keeble, to the effect that, even though (empirically) the object of town planning is primarily physical, town planning's manipulating of the physical environment has social effects.

C3: 'Physical planning has social effects, and so can realise social goals'

This is, simply, an empirical claim, and as such it need not detain us long. It is clearly true that town planning, conceived and practised as an exercise in manipulating the physical environment, has effects on human welfare and, therefore, 'social' effects so defined. Indeed, my résumé (pp. 29–35) of the development of town planning thought since 1945 was, precisely, a résumé of how post-war town planning theorists and practitioners became increasingly aware of the social effects of physical planning. The lesson was therefore learned that, in engaging in physical planning, it was vital for town planners to understand and assess these social effects.

It is by virtue of the fact that physical planning has social effects that its purposes are social, as we saw in our discussion of claim C2 (and specifically C2.1c). Further, it is because town planning has social effects that it is a political practice (indeed, its being political is part of its being social), even though the fact that town planning works primarily on the physical environment has deceived some people into thinking that it is somehow not political, or 'merely technical'. So, C3 is not in dispute.

What *is* disputed about C3 are two issues. First, what *kinds* of social effects physical town planning has, and second, how *strong* these effects are. Again, both these are empirical questions, to be settled by empirical research, and I shall not debate them further here. Suffice to point out that it is in this context that we should situate debates about what is called physical or architectural determinism, i.e. debates about how far the physical form of environments 'determines' social behaviour. I touched on this earlier when describing the social naivety of post-war neighbourhood planning (see pp. 31–32), but it is worth adding that the question of architectural determinism has continued to be a matter of debate in town planning theory, with some (e.g. Newman, 1973; A. Coleman, 1985) taking the view that empirical evidence shows that the form (or design) of the built environment can play a strong role in shaping social behaviour, and others (e.g. Broady, 1968) expressing scepticism at this idea.

Conclusion

This chapter has tried to unpick the various meanings that can be attached to the terms 'physical' and 'social' planning, and the logical relations between them.

As we have seen, there are several different meanings that can be ascribed to claims about town planning being physical and/or social, so my investigation of these various meanings has been unavoidably protracted. Nevertheless, if we now stand back from the minutiae of this analysis, I suggest that two main conclusions emerge. First, the *object* of town planning – at any rate as it has been (since 1947) and currently is constituted in the United Kingdom – is, primarily, the *physical* environment. But second, the *purpose* of town planning is necessarily *social* – 'social', that is, in the sense that the purpose of town planning is the maintenance and enhancement of human welfare. To put this another way, the *means* of town planning are primarily physical, but its *ends* are social. Whether or not the means (or the object) of town planning should be social as well as physical is, as I have also shown, a matter of ideological dispute, and one that has been contested periodically during the history of modern town planning.

These overall conclusions are simple and, I think, unsurprising. They are nonetheless important, for they clarify the scope of town planning as it is currently constituted and, given current political conditions, how it is likely to remain constituted for the forseeable future. And such clarification is needed, for there has been a tendency for some town planning theorists to suggest that town planning can do all manner of 'social and economic' things, and even for them to speak of town planning *being* 'social and economic', as well as physical, planning (see, e.g. Anderson, 1968; Bruton, 1974). To be sure, the *purposes* of town planning are social. They always have been. But what town planning works on to achieve its social goals is the physical environment, as Lewis Keeble understood fifty years ago. The truth of this is therefore a useful reminder of what it is that town planners deal with, and a useful corrective to any town planners who might have pretensions to be doing more than this.

One final observation. Some might say that the conclusion that town planning deals 'only' with the physical environment is rather limiting, rather 'narrow'. I don't see why this should be so, for the quality of the physical environment in which people live is evidently of huge importance to the quality of their lives. Quite how important is not something I can pursue here, but if anyone doubts its significance one need simply point to the numerous campaigns that are fought, sometimes with great passion and sometimes at some risk to campaigner's lives, over proposals to alter physical environments that people have come to love. Or consider the importance people attach to where they live, or where they go for recreation or holidays. Trying to plan well the physical environment in which we live is a major undertaking. As Eric Reade once pointed out, this is task enough for a profession to be getting on with, without having further aspirations to intervene more directly in other aspects of people's social lives, many of which are in any case (as Keeble again implied) the province of other spheres of social policy.

Part II
Groups and issues

4 Gender and rural policy

Jo Little

Gendering planning

The examination of gender inequality in the direction and outcome of planning and policy-making has been almost exclusively confined to the urban context. During the late 1980s and the 1990s an awareness of the differing needs and experiences of men and women infiltrated both the academic analysis and the practice of planning (Little, 1994a; Booth *et al.*, 1996). A greater appreciation of the causes and consequences of unequal gender relations came to inform our understanding of people's daily lives within the built environment and of the effects of the planning process. While such perspectives have been increasingly accepted into mainstream research and practice, they have tended to be spatially concentrated and both the introduction of policies aimed at addressing gender inequality and the study of the relationship between gender difference, planning and the built environment have focused on major towns and cities.

The neglect of gender issues in the practice and study of rural policy and planning is not particularly surprising. As will be elaborated below, the traditional conservatism of many small rural authorities has made them slow to respond to the demands for greater equality within the planning process and much less affected by the kinds of pressures exerted by the new urban left that have been felt by the metropolitan authorities. Research on gender inequality and, in particular, women's needs and experiences within the built environment, has tended to focus more directly on those areas in which change is occurring and where demand for reform in the content and implementation of planning policy has been most concentrated.

The lack of attention devoted to gender and planning in rural areas does not, however, mean that gender difference in such areas has been ignored. A developing interest in rural marginalisation and in the way particular groups are positioned as 'other' in the social construction of rural society has prompted recent research into women's lives in rural areas (Little and Austin, 1996; A. Hughes, 1997). Such work, together with earlier research on women's access to services and employment (Little, 1986) provides an important background to the analysis of policy and to attempts to identify the gendered nature of the construction and outcome of planning and policy-making in rural areas. This chapter uses a number of existing studies on women and rural communities, together with some recent research on rural policy and governance, to provide a broadly based discussion of gender inequality and rural policy from a new perspective. While reference is made to specific findings on, for example, women's participation in local decision-making, the main thrust of the chapter is the discussion of key areas of concern – many of which require further research – and the identification of broad directions in the nature of gender inequality and of existing or potential policy responses.

Gender inequality: women as the rural 'other'

Initial academic interest in rural gender issues derived mainly from the study of farm women. Research identified the differing roles of men and women in agriculture and showed how the work of the farmer's wife was frequently misrepresented within the farm business and almost invariably undervalued. Studies (for example, Whatmore, 1991; Sachs, 1983; Shortall, 1992) demonstrated how women were often involved in many aspects of the farm business, from secretarial work to rearing small animals and providing 'emergency' labour, as well as being (usually) solely responsible for the domestic household. Having identified these patterns of gender difference amongst members of the farm family household, work focused on explanation. Drawing on feminist and Marxist theory, attempts were made to locate women's farm work and their wider domestic roles in the context of gender relations, power and the relationship between production and reproduction. Such work improved both the recognition of gender difference within agriculture and the ways in which such difference related to broader social relations, especially those based on patriarchy.

Following this early interest in farm women, broader studies of gender inequality in rural areas began to emerge. As with the work in an agricultural context, initial emphasis was on identifying and describing the differing needs and experiences of rural women and men. Once the extent and nature of women's inequality had been articulated, attention turned to an examination of the theoretical explanations of gender inequality, particularly the application of feminist perspectives.

The focus on gender difference and inequality in rural areas has led to a much clearer understanding of the difficulties faced by women living in the countryside. Women have been seen to be differentially affected by issues of access and service availability; their domestic role increasing their dependency on local services and reinforcing the problems of poor rural transport. A number of studies have looked, for example, at access to childcare in rural areas (Halliday, 1997; Stone, 1990), showing how inadequate provision of nurseries, playgroups and childminders frequently prevents women from taking up employment and involves them in complex and often costly arrangements revolving around informal support and/or private care. While such studies stress that the problems faced by rural women are not generally *caused* by rurality, they argue that the particular characteristics of rural areas – for example the remoteness and poor service provision – often exaggerate the problems experienced by women in their day-to-day domestic roles. They also draw attention to the ways that the particular experience of rurality (the problems of remoteness, etc.) will vary between women. Factors such as class and age, for example, will have a profound effect on their ability to overcome the more practical/physical difficulties and on their ability to take advantage of some of the perceived benefits of rural life.

Paid work is one area in which the problems of rurality and gender inequality combine to adversely affect women's opportunity. Most rural areas, particularly the remoter areas, are characterised by narrow and restricted labour markets in which the jobs that do exist are frequently low-skill, poorly paid and insecure. Women, because of lower personal mobility and domestic responsibilities, are generally more dependent than men on these local employment opportunities. Consequently, they often have little choice of work and almost no possibilities for career development. Research undertaken in rural Avon in the early 1990s (Little and Austin, 1996), showed that many women had problems finding the sort of paid work that they wanted; 52 per cent of women interviewed as part of a questionnaire survey were employed in jobs that did not reflect their skills, education and training. Their employment histories were fragmented and very disrupted (by childbirth and childrearing, and by job scarcity) and they saw little opportunity for change.

In seeking to explain the inequality of opportunity experienced by women living in rural areas, research on employment has also considered the broader social relations surrounding women's involvement in paid work. Such research has argued that women's (and men's) lives in rural areas are strongly defined according to a set of beliefs and assumptions about rural society. These beliefs and assumptions revolve around the ideal of community and of women's place at the centre of a harmonious and safe society. Davidoff *et al.* (1976) saw the lives of women living in rural communities as tightly bound up in the expectations of family and community; their role as 'linchpins' of rural society shaping both external and internal interpretations and assumptions of their role. The continued power of the 'domestic idyll' as part of the cultural construction of the rural community is an issue that has recently emerged in the examination of contemporary gender relations and social recomposition within rural areas. In debates around difference and marginalisation, the assumptions surrounding women's role in the rural community are seen as influential in defining aspects of women's identities as 'other' to mainstream ideas of rural life.

It has been argued recently that our understanding of rural society and of the lives and experiences of those living in villages needs to acknowledge, to a greater extent, the power and importance of the social and cultural constructions and representations of rurality (Cloke and Little, 1997). The 'rural idyll' has been expounded as the dominant construction of contemporary rurality, representing a rural society of peaceful, safe and healthy communities, largely free of present-day social evils such as crime and homelessness, in which people experience the community as invariably positive, caring and tight-knit. The community of the rural idyll is also characterised by its conservative values and its intolerance of difference. Those groups and individuals not complying with these values are marginalised, seen as 'not belonging' to the true rural community, their views and behaviour as somehow inappropriate to village life and to the wider wellbeing of the countryside. Amongst the groups seen as 'othered' in this way are people of colour, those of non-heterosexual or ambiguous sexuality and New Age Travellers. The otherness of such groups is not only reflected in but also reinforced by the rural idyll – the notion that some are more acceptable than others as rural residents is self-reinforcing as the rural community attracts 'the same' at the expense of 'the other'.

While the identification of groups such as gays and people of colour as other in the context of the rural idyll is relatively straightforward, the nature of personal identity means that the otherness of some groups and individuals may be much more fluid and changeable. Thus it may not be individuals as such who are othered, but aspects of their identity. Women provide a good example, as Cavanagh (1998) has usefully demonstrated. Women as a group of rural residents rarely see themselves, or are seen by others, as marginalised or othered in rural society. They are generally at the heart of the rural community (as noted above) and have a very major influence over behaviour and values within the village. By the same token, however, the dominant assumptions surrounding women's place in the rural family and community renders aspects of their identity as marginal. As I have argued elsewhere (Little, 1997), women as paid employees, especially women in full-time work, are other to the dominant perceptions of the rural woman. The expectation that the 'proper' place of the rural woman is at home with her children, and that women should only take a paid job where this is necessitated by the financial circumstances of the household, frequently results in hostility towards rural women who pursue careers and the marginalisation of those whose paid labour is seen as 'unnecessary'. Despite a recognition of the complexity of notions of identity and of the fluidity of women's marginalisation, little research has been undertaken to establish just what aspects of rural women's identity are othered and in what circumstances. Our understanding of the whole nature of rural marginality is currently very unsophisticated and in need of further exploration.

This very brief review of work on gender and the rural community and on current attempts to situate research on women's lives within the broader context of the cultural construction of rurality has provided a framework for the discussion of women and rural policy. The intention in setting out this framework has been not only to draw attention to existing work on the nature of gender inequality in rural areas but also to show how a focus on women's experiences, whether in relation to paid employment or community involvement, needs to consider the ways in which the expectations and values surrounding women's roles shape their experience of rurality and of social and economic opportunities. The cultural construction of rurality has a very real effect on the material circumstances of women's lives, by sustaining a set of gender identities that serve to marginalise women's economic roles and employment identities. This construction of rurality sees women as located firmly within the family and the community in a way that, as will be discussed below, limits their involvement in public policy, economic regeneration and local governance.

Gender, policy and the new rural governance

The second part of this chapter will now go on to consider gender more directly in the context of planning and the policy-making process in rural areas. As noted above, gender inequality in the experience or outcome of rural policy has received little academic or practical attention. Thus, while a focus on gender in relation to urban policy and planning generally includes an analysis of initiatives that have been introduced to address gender inequality, together with an examination of the (albeit limited) ways power may be shifting to include women's voices, in rural policy gender is something of a non-issue. This having been said, it is worth documenting the lack of attention given to gender in rural policy, since this allows some understanding of the wider directions and assumptions of policy. It also provides a background to the discussion of more recent developments in governance and policy in which different opportunities for addressing gender issues and, in particular, involving women, may be opening up.

Rural women's initiatives?

In the late 1980s and early 1990s two separate pieces of research (Brownill and Halford, 1990 and Halford, 1989; Little, 1994a) revealed the extent of spatial concentration that characterised local authority, especially planning departments', consideration of 'gender issues'. Halford, in a study of 'women's initiatives' in local government, noted that of the thirty-two authorities that had introduced women's committees, sub-committees or an equivalent equal opportunities group, the vast majority (85 per cent) were urban-based authorities – indeed only five authorities outside the London Boroughs or Metropolitan Councils were reported as having a full women's committee or a women's sub-committee (Halford, 1989). The findings of Halford's research were reiterated in later work on women's committees and planning initiatives (Little, 1994a; 1994b). Table 4.1 below shows the decline of formal local authority women's committees and indicates the continued bias in favour of urban areas.

Clearly, the presence or absence of a women's committee is not the only measure (or necessarily a particularly accurate one) of a local authority's attitude to gender issues. The existence of a women's committee does, however, indicate a certain recognition of the particular needs of women and of the importance of direct action to improve women's position and reduce gender inequality. It has

Table 4.1 Local authority women's committees in England, 1989, 1994

Women's committees	1989	1994
Full committee	14	12
Sub-committee	8	5

been shown, moreover, that, in terms of planning at least, those authorities where Women's Committees had been established were more likely to have introduced policies aimed at women than those where no women's committee existed (Little, 1994a).

Focusing specifically on planning can provide a more sensitive assessment of authorities' commitment to reducing gender inequality. In 1991 a survey of all local authority planning departments in England allowed the existence of policies on gender and women's initiatives to be identified more precisely and also provided an opportunity to look behind the occurrence of policy at the attitudes of those involved in the planning and policy-making process. The survey confirmed that the distribution of women's initiatives in planning mirrored that of Women's Committees in terms of the bias towards urban localities. Taking all women's and equal opportunities initiatives it was found that while 57 per cent of all London boroughs and metropolitan councils had introduced some form of policy, only 13 per cent of districts had done so. In addition, where London and metropolitan authorities had introduced measures aimed at addressing gender inequality they had tended to be part of a package including a range of different initiatives through different policy areas. In the districts, any measures were generally more limited and small scale, addressing one particular 'problem' or policy area in which a very specific need was identified and targeted.

Explaining their lack of women's initiatives, some of the more rural planning authorities drew attention to their size, claiming that they were too small to get involved in more peripheral planning issues such as gender inequality. There appeared to be an assumption in the comments of some that gender issues were essentially *urban* issues and that initiatives were to address *urban* problems that were not as important (or simply not applicable) to rural areas. As one questionnaire respondent claimed: 'There is no specific attention given to women's issues. The department is small and the district is rural.'

The absence of initiatives aimed at addressing gender inequality in rural areas was also put down to a lack of pressure or support for such measures from local populations. Rural areas rarely exhibit the sorts of locally based women's pressure groups seen in urban areas. Indeed, the rise in women's groups in the early 1980s was a phenomenon very firmly associated with the New Urban Left (Little, 1994b) and with renewed support for a left wing political agenda. As Brownill and Halford (1990) have pointed out, the presence of grassroots political movements are critical to the establishment and success of local authority women's committees and to a commitment to equal opportunity measures. In rural areas the local political will to push for initiatives aimed at addressing gender inequality rarely exists in any coherent form.

The lack of commitment to women's needs in rural areas is also clear in other policy sectors. In 1991, a review of Rural Development Programmes (RDPs) (Little, 1991) examined the extent to which policies for rural regeneration as formulated and implemented by the Rural Development Commission (RDC) incorporated any particular recognition of women's needs or gender inequality. The review found that, of the twenty RDPs analysed, twelve noted the 'problem' of low activity rates amongst rural women and the lack of suitable local employment. Typical of such recognition was the North Yorkshire RDP which stated:

> Low density of population, under-employment, long journeys to work, lack of readily available training and re-training opportunities, *limited opportunities for female employment*, lack of job variety, entry into routine jobs below individual ability, out-migration of young people and others of working age, imbalances in age structure of the population, low income levels, a steady decline in the quality and range of services and a heavy reliance on grants and subsidies.
>
> (Yorkshire RDP, 1985, p. 1, quoted in Little, 1991; emphasis added)

Having listed poor employment opportunities for women amongst the economic problems faced by rural localities, however, practically all the RDPs failed to make further mention of women's needs; only three Rural Development Areas, Gloucestershire, Northampton and Somerset, followed up the recognition of a problem with any specific policy objective on women's employment. Thus the Gloucestershire strategy stated:

> there is a pressing need to improve job opportunities in the area for both men and women and in particular:
> to reduce the very high levels of unemployment
> to widen the range of job opportunities
> to stimulate growth in the service sector which is poorly represented in the area
> to provide job opportunities for women and school leavers who make up large components of the unemployed.
>
> (Gloucestershire RDP, 1985, p. 2, quoted in Little, 1991)

Even stating, as a policy objective, the need to address the problems surrounding women's low activity rates failed to lead these areas into committing themselves to any particular initiatives. Indeed, out of all twenty RDPs only one included any form of initiative aimed at women's needs and this was a leisure initiative introduced by Cumbria RDA.

New rural governance

It has been suggested (M. Goodwin, 1998; Jones and Little, forthcoming) that rural areas, like urban areas, are undergoing profound changes in the way in which they are governed. A system of government is being replaced by one of governance whereby, according to Goodwin:

> government signals a concern with the formal institutions and structures of the state, [while] the concept of governance is broader and draws attention

to the ways in which governmental and non-governmental organisations work together, and to the ways in which political power is distributed, both internal and external to the state.

(M. Goodwin, 1998, p. 2)

This new concept of governance has its origins, according to many theorists (Jessop, 1995; Tickell and Peck, 1995) in the shift from a Fordist to a post-Fordist system of capital accumulation and in the associated changes in the mode of social regulation. As well as including new agencies and organisations in the decision-making process, new forms of governance have prompted a change in the culture of policy making and in associated social and economic structures. There is insufficient time for a lengthy discussion of these theoretical issues – the key question here is whether the sorts of changes taking place as part of the emergence of a new rural governance incorporate or imply any implications for the treatment of gender inequality and/or women's needs.

The next section of this chapter, therefore, will start to focus on whether the new structures and practices of governance – the 'tangled hierarchies' (Jessop, 1995, p. 310, quoted in M. Goodwin, 1998) that increasingly govern rural areas – and the new cultures of decision-making suggest any change from traditional policy directions in terms of gender inequality, or offer any new opportunities for women's greater involvement in the rural decision-making process. It considers in particular the claim that new forms of rural governance enable a greater range of people to become involved in policy in the move towards partnership and 'active citizenship' within the broader operation of the mechanisms and procedure of the local state and associated agencies.

After a few observations on rural local governance in general, discussion here will focus largely on one particular example of rural policy, a regeneration initiative called Rural Challenge, which exhibits some of the tendencies associated with the shift from government to governance in the formulation and outcome of rural decision-making. Focusing on one initiative in this way provides an opportunity to look in some depth at the construction and operation of policy at the local level and to learn more about the negotiation of particular decisions amongst local agencies and community members.

Tickell and Peck (1997), in a study of local politics and economic policy in Manchester, consider what they term the 'regendering' of local governance. Drawing on theoretical discussions of gender and the local state, they examine the representation of women in the agencies of local governance and in the power relations surrounding economic policy. They argue that the shift to new forms of governance has meant a change in the balance of local political power in favour of the private sector and quangos, and that such a change has resulted in the emergence of a new local business elite with an important role in shaping the direction of policy and funding. This elite, Tickell and Peck argue, is dominated, both in terms of its composition and its culture, by men. Thus they write that:

Manchester's new business elite are . . . the 'new Manchester Men'. [This] reflects both the new found self confidence of the business community and the implicit power of its reconstituted and male-dominated elite networks. But it is not just in the sexual composition of the new governance institutions that local politics have become (re)masculinised in Manchester; their favoured discourses and their *modus operandi* also serve effectively to exclude women from the political process.

(Tickell and Peck, 1997, p. 607)

According to Tickell and Peck this masculine culture or *modus operandi* which excludes women is manifest in the importance of male networks, the language of meetings and an essentialist attitude towards 'female qualities' in the business world. While the new political elite in Manchester is 'not one which actively rejects female participation it is one where women are not considered as significant political players' (Tickell and Peck, 1997, p. 608). Interestingly, as private business becomes more critical of the state as the provider of services it is described in female terms – classically the 'nanny state' which is contrasted with the hard, male world of business. As far as women's contribution within new forms of governance is concerned, this appears very much restricted to the voluntary sector and to community-based action, both of which have been relegated to a secondary role in relation to the private, business-led sector.

The concern here is whether the emergence of a new rural governance has led to similar tendencies as those outlined by Tickell and Peck in relation to Manchester, in terms of the cultures of decision-making and the domination of a particular set of private sector business (masculine) interests in rural areas. As noted above, with greater opportunity being supposedly created for the involvement of local communities and active citizens as partners in local decision-making, the expectation would be for greater involvement of women and a broader base for rural policy. The following brief discussion of the Rural Challenge initiative begins to test some of these assumptions.

Rural Challenge and gender

Rural Challenge (RC) is an initiative introduced by the Rural Development Commission, designed to stimulate economic and social regeneration in rural communities in England. The scheme awards prizes of up to £1 million annually to projects put together by partnerships of local public, private, voluntary and community interests. Six prizes are awarded each year and projects are selected by a series of staged competitions (in the manner of the former City Challenge) in which teams have to present their schemes to panels of judges at both the local and national level. There is a requirement that all the sectors are represented in the bid and that the schemes chosen have the support of the local community. It is also a requirement that the bids indicate matched funding opportunities, making the final size of most bids in excess of £2 million. RC is currently in its fourth year of operation and has met with mixed reactions from communities and policy-makers (see Little *et al.*, 1998 for a fuller discussion).

In its design and implementation RC does demonstrate, as indicated above, some of the characteristics of a new, post-Fordist governance. The emphasis in the policy is on the development of partnerships for the delivery of rural regeneration and, in particular, the involvement of the private sector. Projects are required to be innovative and entrepreneurial, competitive and flexible; running through the policy documentation is an assumption that the 'hard-nosed' attitude of 'business' and the private sector will be more successful in promoting and sustaining rural regeneration (RDC, 1995). While the requirements for community involvement imply a broad base for decision-making, the expectations of the form of community participation are limited and revolve mainly around social and service provision.

As in the urban situation described by Tickell and Peck (1997), the entrepreneurial culture of RC and the emphasis on the private sector had resulted in the majority of initiatives being very masculine in terms of both content and management. Most RC schemes have at their centre a major capital development – a bricks-and-mortar project – designed to generate income and provide employment. The emphasis is on the needs of business and these are assumed to be the needs of the local community. None of the schemes include any special emphasis on women's needs, either in terms of business development or employment provision. The schemes are often aimed at the creation of local employment to replace previously existing (male) jobs, in a very partial interpretation of local economic/labour market histories. Hence the perceived need in a Leicestershire RC scheme to replace the jobs lost through the closure of local coal mines and the need, in Somerset, to replace jobs lost through the decline of the fishing industry.

In most of the RC projects the involvement of the community is seen as something of an 'add-on' feature. While the rules of the scheme allow for any sector, public, private, voluntary or community, to take the lead, in only one project of the first eighteen to be funded is the lead partner a community group (with a second being led by the local Rural Community Council, a voluntary agency). In terms of individual schemes, priority in terms of funding is generally with the business-led, job-creation elements with the more service, educational and training needs of the local communities taking second place. Indeed, it was admitted by some of the project officers in discussion that the community element of their RC scheme had been imported from existing work for the purposes of the wider bid. In other cases, the views of the community do not always coincide with the broader objectives being proposed by the RC scheme, so their involvement tends to be sidelined. While one would not wish to fall into an essentialist argument which sees women's interests as simply relating to community services/provision, it was clear that it was in this area that they were most likely to become involved as 'active citizens' in the decision-making process. By downplaying this element of RC women's voices were marginalised.

It is not only the content but the management of RC schemes that was found to be masculine in orientation – again reflecting the male culture of the new forms of rural governance. The personnel involved in the formal running of the scheme are mainly men – of the eighteen project officers interviewed only one was a woman. Again, the association with capital development and big projects served, it was thought, to marginalise women in terms of the formal roles of project management. As one project officer stated:

> The big capital projects are not necessarily male specific, but certainly not within the realms of experience of a lot of women – and maybe not even a very attractive job.
>
> (Project officer, RC scheme)

while another noted:

> My background is that I'm a chartered secretary and a chartered builder and other things as well and that stands me in good stead for servicing the [RC] board meetings and everything.
>
> (Project officer, RC scheme)

The male ethos or culture of the RC scheme is perpetuated by the ways of working. One project officer, for example, spoke of the formality of working practices in his case as potentially off-putting to many women: 'it's all grey suits and proper agendas and that sort of thing'. In another case, however, it was an *informality* of approach that allowed the project to be dominated by a 'male club' type of attitude:

> Even at meetings it's very cultural – 'do you want to be on Harry?' 'Do you want to be on George?' So I put my hand up – 'Excuse me, don't you think there should be a woman representative'? Women get overlooked. As I say, it's a cultural thing.
>
> (Project officer, RC scheme)

This masculine culture is also significant in the more general underpinnings of the RC scheme itself. The idea of staged competition as a means of allocating prize money is seen as confrontational and unlikely to lead to co-operation between project deliverers. Project representatives who had been involved in the bidding process described it as very intimidating and hostile. One woman project leader whose project was turned down recalled the national selection panel as 'absolutely awful'. Flashy projects presented by men in sharp suits were seen to have a far better chance of selection than worthy initiatives put together and delivered by local community activists. The gender associations and implications here are implicit rather than explicit but, nevertheless, very real. The male ethos within which RC is constructed and run is also felt to have its roots in the masculine culture of the RDC itself. While recent restructuring of the organisation has placed some women in regional management positions, it is still perceived as highly male dominated, especially further up the hierarchy. As one project officer put it:

> I was involved in a stall in the RDC's arena at the [Great Yorkshire] Show. There must have been twenty or thirty RDC suited characters – all male – wandering around, just out for a jolly.
>
> (Project officer, RC scheme)

Conclusion

This chapter has provided a very broad-ranging account of women and policy in contemporary (UK) rural areas. It has established the state of current work on women and rural policy and on the understanding, theoretical and empirical, of gender inequality in rural communities. Inevitably, much has had to remain unsaid in this review because of the space restrictions and the need to move on from existing arguments. In attempting to move on, then, the chapter has argued that new efforts to comprehend gender inequality and the rural policy process need to be located within two recent areas of theoretical debate which have emerged in the context of work on rural social and economic restructuring. These are, first, the cultural construction of rurality and the implications for rural marginality and, second, the shift from a Fordist to a post-Fordist mode of social regulation and the emergence of a new rural governance.

The first of these areas provides a new perspective on women's marginality in rural areas, helping to recognise that women's needs relate to complex and fluid notions of gender identity. While all women are marginalised in certain respects by male power and the operation of gender relations, the particular manifestation of that marginalisation in rural areas is closely linked to constructions of rurality. So it has been argued here that women's lives and experiences are shaped by the expectations surrounding *rural* gender identities, expectations that may not marginalise women as a group from rural society, but which serve to 'other' parts of their gendered identities. An understanding of the values and assumptions surrounding women's lives in rural areas – their role in the community and family and their lack of involvement in formal politics and the economy – is crucial to the broader investigation of women's experiences in relation to, for example, paid work or service provision.

The second area on which this chapter has focused is the emerging political framework of the new rural governance. It has been suggested that rural areas are experiencing a shift away from traditional forms of government towards a system of governance in which new structures and processes have emerged and whereby the policy world is now made up of 'diverse, overlapping and integrated networks, often operating beyond effective control by formal structures of government' (M. Goodwin, 1998, p. 5). The mechanisms and the relationships that are emerging in rural areas have, it has been argued here, important implications for both the content and culture of policy. In the example used in this chapter, that of Rural

Challenge, it was argued that aspects of the content and culture of emerging rural policy are highly masculine and serve to underplay or marginalise women's needs. Thus the gender implications incorporated in the new rural governance may be profound. As yet this is a topic that has received little attention. If the arguments put forward here are to be accepted then it is important that the debates surrounding the direction and outcome of new forms of rural governance incorporate a specific gender dimension.

Both aspects of the chapter have sought to demonstrate the importance of space in the social construction of gender roles and gender relations. Attention has focused on the particular characteristics of rural spaces and places, arguing that it is possible to identify a definite *rural* influence on the operation of gender relations and on the varying experiences of men and women. Care must be taken not to attribute causal powers to the notion of rurality – women are not disadvantaged because they are rural women. It is important, however, that we recognise the difference that space makes in the way gender relations are played out, whether that is a physical/practical difference affecting, for example access to services or the provision of basic facilities, or whether it is a cultural difference affecting expectations and assumptions about rural values and lifestyles (or both). The emphasis of this chapter has been on the interrelatedness of these factors and on the spatial configuration of resulting social processes and patterns. Future work on the lives of men and women living in rural communities and on the operation of gender relations in the countryside needs to keep such a perspective in focus.

5 Planning to grow old

Rose Gilroy

Introduction

In the coming decades, urban and rural planners will be confronted by a new challenge. The developing world is ageing. Not only is there a greater percentage of older people in the population because of falling birth rates but more people are living into their ninth or tenth decade. This trend has implications at the community level, where the demands for housing adaptations or housing alternatives, different modes of transportation and services will need to be addressed. The ability of communities to respond to these needs will be influenced by how well the concerns for older adults are integrated with planning efforts.

Older people: diversities and commonalities

Throughout this discussion the term 'older people' rather than 'old people' is used. This is in line with the terminology used by the European Year of Older People and Solidarity Between Generations (Gilroy and Castle, 1995). The term acknowledges our own participation in the universal ageing process. If it is agreed that ageism not only uses chronological age as a means of marking out a class of people who suffer the consequences of this oppression, but also generates a fear of the ageing process, then it is damaging to suggest that those who discuss ageism are somehow exempt from the ageing experience (Bytheway and Johnson, 1990). The phrase suggests a continuum of life and experience as opposed to one in which a person crosses a line after sixty or sixty-five birthdays to become 'the other'. This is an important point. If we see change and changing need as a continuum rather than a form of deviance we then conceptualise the ageing society as presenting challenges and opportunities for planners and urban/rural policy-makers to change and accommodate diversity. Later in the discussion we consider how modernist thinking has led planners and policy-makers to 'put old people in their place' by creating age-segregated environments, and how this is being consciously challenged in various parts of the world.

The phrase 'older people' is a movement away from modernist thinking which suggests that at post-retirement there is homogeneity created solely by chronology, although, for example, between those aged 60 and those 90 years old there are clear generation gaps, with different life experiences and shaping influences. In addition, as throughout existence, gender, class, income and ethnicity intersect in varying degrees to define life (Beall, 1997). It is to these issues that we turn next in examining the data on older people.

Demographic trends

Living to a great age is a twentieth-century, even a post-Second-World-War, phenomenon. In Britain current figures show that people aged 55 and over now make up a quarter of all citizens and that this will increase to more than one-third, with the number of people of working age falling. These trends can be seen replicated across Europe. In 2025 the number of young people aged under 20 will have fallen by 10.6 per cent (9.4 million) and the number of adults currently considered to be of working age (the groups aged 20–59) will have also fallen by 6.4 per cent (13.2 million) compared to 1995. Over the same period as younger people decline in number the number of adults aged 60 and over will increase by 48.7 per cent, that is 37 million persons (Social Europe, 1996).

A critical factor which demands some reflection is the dominance of women. Sixty-five per cent of the retired population is female, while among the oldest group (those aged 85 or more) 75 per cent are women, as shown in Table 5.1. As in discussions of disability, gender aspects of ageing are often understated, which may bear out the statement by Szinovacs that

> women's retirement has been neglected because a major stumbling block was the belief that it did not constitute a salient social or research issue, backed up by the highly questionable but prevalent assumption that retirement from work is a less significant event in women's lives.
>
> (Szinovacs, 1982, quoted in Ford and Sinclair, 1987, pp. 2–3)

Table 5.1 Population by age and gender (percentages)

	Under 16	16–34	35–54	55–64	65–74	75 and over
Mid-year estimates						
1995	21	27	26	10	9	7
Males	22	29	27	10	8	5
Females	20	26	26	10	9	9
Mid-year projections						
2001	20	25	29	10	8	7
2011	18	24	29	12	9	7
2021	18	23	26	14	11	8
Males	18	24	26	14	10	7
Females	17	23	25	14	11	10

Source: ONS (1997), p. 39, Table 1.5.

The increased participation of women in the labour market and the growing number of women who see their work as a career rather than a job may alter this. It may be that women's dual role as workers and domestic workers/carers results in women's concept of themselves resting on more than the single support of waged work (Gilroy, 1996a) and therefore retirement from the first does not produce the trauma that it often triggers in the lives of men. Of course this is conjecture: what is missing from the discussion is the experiences of older women. Ford and Sinclair (1987) assert that not only is their voice absent but that it is thought that 'their experiences can be understood without recourse to investigation, either because it is assumed that the pattern of their earlier life is continued into old age or that the findings from studies of men in old age can be extended to cover women' (p. 2). A challenge for planners and researchers is to find effective ways of bringing these voices into the debate and into the academy. This will be developed later in the chapter.

The feminisation of later life has economic consequences too. Most EC pensions are related to income and working years. So, as Ilona Ostner says (1993), the widows of low-income workers risk poverty even when they have been hard-working wives. For women who live without men the problems of lower earnings, broken employment records for periods of caring and part-time work that may have prevented building up a contribution record all militate against a decent income and the ability to make choices in later life. The changing face of pensions in Britain, with a greater emphasis on occupational pensions, has been heralded as a turning point in reducing the risk of poverty, but without a revolution in the working patterns of women there will be no radical change. The feminisation of poverty in old age is a problem that will not disappear; in fact it may increase with the probable increase in the divorce rate. Harper and Laws (1995) discuss divorce as a component of an ageing society, suggesting that the institution of marriage needs some refurbishment if it is to cope with forty to fifty years of cohabitation, when it was established to last for far shorter periods.

The dominance of women also has implications for governance and involvement. A number of studies have explored the role of women in protest movements and in community development but few British studies have mentioned the role of older women (Wood *et al.*, 1995). Is this because there are few examples or does it suggest ageist attitudes among researchers? Harper and Laws (1995) contend that women become more assertive as they age, while men become more dependent and expressive – which overturns the stereotype of the helpless old lady and substitutes the confident woman who knows the system and will not be browbeaten. Far from being passive members of the community, older women may be seen as a resource in galvanising neighbours to take action. They have the time, the experience and the vim. What is the cost, therefore, of taking older women out of their neighbourhoods and rehousing them into age-segregated communities?

Table 5.2 Population by ethnic group and age, spring 1996 (percentages)

	Under 16	*16–34*	*35–54*	*55 and over*
White	20	27	27	26
Black Caribbean	23	36	24	17
Black African	28	43	23	6
Other Black	49	38	12	—
Indian	27	32	29	12
Pakistani	40	35	17	8
Bangladesh	40	35	17	8
Chinese	16	40	30	15
Other Asian	27	31	36	6
Other minorities	51	30	15	5

Source: ONS (1997), Table 1.10.

Ethnicity

Britain is a multiracial society, but the percentage of older people from ethnic minority communities remains small. For example, only 12 per cent of the Indian population of 877,000 is aged 55 and over, while the Afro-Caribbean community has 17 per cent in this age group from a total population of below half a million (see Table 5.2). In all the minority communities, with the exception of the Chinese, the number of under-16s outstrips the number of older people. Nevertheless, it will be important for policy-makers and service deliverers to get into dialogue with older people from all communities to determine needs and aspirations.

Disability

While disability has a rising political profile, and we finally have a more comprehensive legislative instrument in the Disability Discrimination Act of 1995, data on

Table 5.3 Age distribution of disabled adults

Age group	*Disabled adults*		
	Men	*Women*	*Total*
60–4	14	8	11
65–9	13	11	12
70–4	14	14	14
75–9	12	15	14
80–4	8	13	11
85 and over	4	10	7
All 16–59	35	29	31
All 60–74	41	33	37
All 75 and over	24	38	32

Source: Martin *et al.* (1988), p. 6, Table 2.1.

the numbers of disabled people remains poor. The 1991 census included a question on limiting long-term illness, which has produced a number of new data sets but these do not answer our questions on disability and disabled people. A person who is blind is not ill. The discussion here relies unsatisfactorily on data collected in the 1980s, which indicates that older people are more likely to be disabled and that this has a strong gender dimension (see Table 5.3).

Perhaps it is another aspect of the negative approach to old age, but it is frequently assumed that disability in later life is a once-and-for-all condition. R. Coleman (1992) suggests that the medical model of disability has become entwined with a medical model of ageing, such that older people are seen as naturally disabled by their age when the real problem is that society has failed to pay attention to their needs. Disability, at least in the ability to live an independent life, is tightly bound up with environmental issues and over these issues planners exercise control and thus have responsibility. The next section, therefore, considers older people specifically.

Planners and older people

British planners looking to respond to the ageing society have been left very much on their own. There are scant mentions in PPGs (Planning Policy Guidance Notes produced by the DETR) to remind planners that older people use the built and created environment too, and that attention should be made to their needs. Members of the Royal Town Planning Institute are not reminded to consider the impact of ageism since age is excluded from their code of conduct. The omission here means that there is no Planning Advice Note, nor any hope of one, to alert planners to the possible concerns of older people and to consider ways of bringing their voices into the debate. All this suggests that planners, planning and older people have so few points of contact that it is not worth giving guidance.

Indeed, when we consider those services that have greatest impact on the lives of older people, we tend to turn to housing, to social services and to the health service. These are the recognised key players in community care which, when applied to older people, is concerned with keeping the vulnerable in the community by delivering care packages in their own homes or appropriate domestic-scale settings, which might include sheltered housing or residential care homes. An examination of articles on planning and older people in recent years demonstrates, in the main, a lack of awareness of the place that older people have in the everyday life of the neighbourhood and an over-concentration on issues surrounding the siting of nursing homes. Since the majority of older people will never need specialist provision of any kind (Laws, 1994) there is an assumption that planners only touch the lives of the very old and frail. Leather asserts:

> Planners, it seems to me, have treated older people essentially as a residual or even a problem group in the past, and in the future I think it must be

recognised that they will be a very strong lobby and also a powerful force for demand, not one which can be expected to be satisfied any longer with poor environments.

(Leather, 1997, p. 4)

The later part of life has been constructed therefore as a problem, whose group members must adapt to and fit in with as best they can or 'have their needs met and catered for as victims' (Beall, 1997, p. 16) or as members of a vulnerable group. A more holistic view of the relationship between planners and older people was set out in the 1980s:

Since the 1947 Town and Country Planning Act, local government has had a clearly defined responsibility for planning and managing the environment for the benefit of the community as a whole. The extent to which planners have seen older people as members of that community must be questioned in the light of some planning decisions which have placed less affluent and less mobile people at a disadvantage. Many city centres are 'no-go' areas for the less athletic who cannot dash through fast moving traffic, nor manage pedestrian subways (often intimidating places anyway); similarly the lack of places to sit and rest with shopping bags, and the absence of conveniently sited public toilets can make city centres an ordeal – and not just for the older generation . . . Balancing the needs of all members of a community is no easy task, but an awareness of the limitations and requirements of older people is crucial to sensitive planning and will in many cases benefit everyone else who lacks private transport or whose mobility is restricted, whether by a sprained ankle or baby buggy.

(Norton *et al.*, 1986, p. 2)

The friction between the less mobile and the fast pace of urban life is usually framed within discussions of disability, while the majority of those who bear the full brunt of this are older people. This focus on the mismatch between urban life and the ageing of society is an issue which has preoccupied Japanese planners (see Box 5.1). The Japanese concern to interpret an ageing society as an issue that impacts on the whole built environment and services is a laudable model for all planners in the developed world to follow. A more common model has been to see older people as a special-needs group whose lives are bounded by activities and concerns that are very different from ours (the non-elderly). The little research that has been undertaken (Churchill and Everitt, 1996) illustrates that, regarding older people's social activities and the development of programmes to enhance them, there is little different about being old in terms of interest and aspirations. What makes the difference is often the lack of accessibility, making it difficult to participate because of such things as cost, transport, ageism and fear of crime. Planners might begin by asking older people what quality of life means to them and exploring the barriers that might prevent their access. This would at least provide the beginning of a dialogue with older people and might present an agenda

Box 5.1 **Ageing in Japan**

In Japan the ageing society is seen as an enormously important issue. Between 1970 and 1990 the percentage of over 65s in the population doubled and is set to double again by 2020. The Japanese approach has not been to segregate older people but to adapt cities to meet the needs of the ageing community. Between 1973 and 1989 more than three hundred projects were launched across Japan by the Ministry of Health and Welfare all geared to creating what has come to be known as 'the gentle city'. A project by the Ministry of Construction in 1993 aimed to widen pavements, install pedestrian overpasses with escalators and to improve the access to public transport. A further initiative by the same ministry entitled 'Creating space for welfare' creates an impetus to draw up guidelines which focus on the needs of older people, those with disabilities and parents with young children.

In 1993 a study was carried out in Tokyo to identify the necessary measures that would need to be taken to create an urban environment which aids mobility and makes travel more comfortable. The study revealed that everyday commuting involved an expenditure of energy equivalent to walking up the stairs of a six-storey building and even younger people spoke of the experience as tiring. There were too few energy savers in the shape of seats, escalators and ramps. Displays of information were in too small a character to be read easily by the weak-sighted.

(Matsukawa, 1994)

for action. Planners might ask in what way they can affect some of these issues, such as affordability. It might not be possible to influence the cost of entrance to the cinema, but using planning gain to provide transport links between neighbourhoods and supermarkets that offer good value might ensure that an older person had more money in his or her pocket. Similarly, putting weight behind energy efficiency schemes; linking with Age Concern's Winter Action campaign and promoting loft insulation and draught-proofing can mean that less money is spent on essentials and more on enjoyment.

Housing

For British planners it seems that the major issue raised by the ageing population is the question of where these people will live and what type of dwellings they will occupy. Again, how this is conceptualised determines the solution. Evidence from a DoE research project contests the view that all older people are living in inappropriate housing and that all are in need of support:

Sixty-six per cent of households headed by an older person have no assessed need for specialist housing or other housing with care support to remain at home.

Sixty-eight per cent of these households have no wish to move and are not physically or mentally dependent.

Eighteen per cent are physically or mentally dependent, but do not wish to move and state that they have sufficient formal and/or informal support and/or aids and adaptations.

Fourteen per cent would fail the current means test for receipt of grant funding towards repairs and adaptations to their homes.

(Adapted from McCafferty, 1994, p. 4)

A study in 1995 of six English planning authorities with higher than average concentrations of older people revealed that housing for older people was seen as specialist housing. None of the plans discussed the concept of lifetime homes and only one discussed the need for care and repair which would help older people to adapt their current dwelling rather than be disrupted by moving house (Gilroy and Castle, 1995). The proposed extension of Part M of the Building Regulations to housing will, eventually, create homes in which we might be able to live for the whole of our lives with greater ease. However, it has taken a long time for policy-makers to realise the psychological importance of home and the fact that it is more efficient to help people 'stay put'.

Laws (1994) sees the growth of age-segregated solutions such as sheltered housing and nursing homes and the subsequent spatial separation between generations as a factor contributing to the social separation of older people. The people and professionals of Toronto felt strongly enough about this when thinking out the vision for the future of their city to assert that:

> We should not provide separate facilities for older people but try to integrate older people into the mainstream of Canadian society. Seniors should be housed in mixed communities and buildings should be designed so that they are accessible to people of all ages . . . there should be no reference to housing segregated by age in the official plan.

(PDD, 1991, p. i)

Similarly, in the participative process that resulted in the drawing up of Vancouver's CityPlan (Gilroy, 1996b) there was an unequivocal demand from the neighbourhood workshops (called City Circles) for multi-generational communities:

> Ideally, it should be possible for people, if they so wish, to live in their own neighbourhood. The phases of a lifetime – growing up in a neighbourhood, then as young adults, rearing children, becoming empty-nesters and eventually retiring as seniors – all these different phases suit different housing types and prices.

(The Eighth City Circle, 1993 as discussed in Gilroy, 1996b)

In the final CityPlan (City of Vancouver, 1995) these concerns were incorporated into a vision of the city with greater intensification of dwellings and a greater emphasis on neighbourhood centres where people might find shops, jobs, neighbourhood-based services, safe public places and a place to meet with neighbours and join in community life:

People will have more opportunities to live in their own neighbourhoods as they pass through various ages and stages of their lives. More housing will be available in neighbourhood centres to allow older and younger people to remain in their familiar neighbourhoods as their needs change.

(City of Vancouver, 1995, p. 12)

We might assert that in the UK this separation has only been slight, really a feature of US developments rather than the British experience. We have no Sun City developments, though some local authorities have explored similar, if smaller, models (see Box 5.2). While purpose-built communities are rare in Britain, most major settlements have their share of sheltered housing which, it is said, is part of the community but whose built form marks it out as a different. Sheltered housing has had its champions and its critics: their arguments are summed up in Box 5.3.

Box 5.2 Ageing in Newcastle

In the early 1980s Newcastle Metropolitan Borough Council was actively exploring the concept of 'a village for the elderly' which would include a sheltered housing scheme, a range of shops and facilities, a residential care home, a home for mentally infirm people, hospital facilities for those with physical and psycho-geriatric problems and a mortuary. The scheme had many proponents who spoke of economies of service delivery. How it might feel to live on a conveyor belt was rarely debated. The scheme finally foundered on the drawing board because of financial issues. In the process of drawing up and considering the range of facilities available there was no effort made to determine the views of older people themselves.

(Author's experience)

Those who argue the need for sheltered housing on the grounds of the importance of good housing for health and wellbeing may consider how much the problem has been exacerbated by the failure to invest in existing housing stock. The paucity of local authority new-build and the reduced new build programme of housing associations has reduced the number of small flats and small houses which might increase the range of choices for older people. In addition, there is a poor policy response to the majority of older people who are owner occupiers and whose dwellings may also be ageing or present difficulties. Together, these result in limited choices for older people who may apply to sheltered housing (rented or for sale) because it is their only opportunity of getting a well-located, warm dwelling. Churchill and Everitt (1996) were told by sheltered housing tenants that the major benefit in moving was feeling safe, but that this was gained at the cost of feelings of independence and control:

Without the crime, the burglaries and the vandalism, I would have stayed in my own home. It was lovely. Right on the Terrace. I'm safer here.

(Churchill and Everitt, 1996, p. 31)

Box 5.3 **Sheltered housing issues**

Sheltered housing provides better-quality housing
Sheltered housing provides more than housing. Why fill it with those who don't need the extra facilities?
   More cost-effective to improve existing dwellings.
   Cheaper to build small flats than sheltered housing with its associated revenue costs.

Sheltered housing provides for special needs
An ageist response in separating people on age grounds.
   Glosses over other needs among individuals.

Sheltered housing helps retain independence
No evidence that it empowers or helps people to develop or exercise their capacities.
   For those who do not need the support offered it may accelerate dependency.

Sheltered housing provides a safeguard in emergencies
Categorises older people as vulnerable.
   No data available on use of call alarm or reliance on the warden as a source of support.

Sheltered housing gives choice vis-à-vis residential care
No evidence that older people have information to make a real, informed choice.

Sheltered housing counters loneliness
No evidence that being in sheltered housing creates a new social life for tenants or owners.

In addition to these small age-segregated communities, we do have the phenomenon of retiring to the seaside, such that more than 30 per cent of the population of a town such as Worthing is over retirement age (Gilroy and Castle, 1995). For the future, Leather predicts that the next group of retirees 'may be the most affluent generation in old age that we may ever see' (1977, p. 1), because most will be home owners (see Table 5.4) and many will have higher incomes. Those who follow may be less affluent, because they will have been affected by the casualisation of the job market. There will, of course, be many who have low incomes and are constrained in their choices by lack of income and disposable assets. However, those who can choose a new location in retirement may, as Leather predicts, seek rural locations, not remote villages but market towns:

> small towns and large villages with access to the services that are needed and valued by older people: a grocery shop, the chemist's, the doctor's. . . . Those are the services which people need on a frequent and daily basis. They

Table 5.4 Tenure profile of heads of household by age in Great Britain (percentages)

Ages at 1995	Owners		Renters				With job	All tenures
	Owned outright	With mortgage	LA	HA	Private unfurnished	Private furnished		
Under 25	0	2	5	6	12	28	5	4
25–9	0	10	7	12	14	24	15	8
30–44	5	46	24	27	23	32	42	29
45–59	23	33	18	16	16	9	24	25
60–9	15	6	16	13	11	2	12	16
70–9	44	2	19	17	13	3	2	13
80 or over	13	0	10	10	12	1	0	6

Source: Wilcox (1998), p. 106, Table 28a.
Notes
LA   Local authority
HA   Housing association

will try, if they can, to move to places where they can easily and safely get to those facilities.

(Leather 1997, p. 2)

As Leather states, discussions of the projected 4.4 million extra households by 2016 (e.g. DoE, 1995a) seem to have omitted reference to older people, but their invasion of the small towns and large villages will price out younger and less affluent people. Have we fully thought out this issue?

Participation

One of the main thrusts of the thinking behind this discussion is the lack of dialogue between older people and professionals on matters which affect the place in which they live or concerning the quality and frequency of services specifically offered to them. This section examines some of the barriers to involvement for older people and considers some of the innovative ways in which older people have been included.

Since the Skeffington Report in 1968, it has been recognised that many groups are unable to participate because of inequalities which could, in part, be ameliorated by planners and other professionals taking a more proactive stance in seeking their involvement. Any examination needs to consider the interrelationship between equity, access and resources. The ability and extent of an individual's capacity to participate will depend largely on resources. These resources may be individual (income and education) as well as organisational (links to groups, to networks of support). Lying behind these resources are social, cultural, ideological and economic factors, all of which have a bearing on equity and increase or diminish the ability to obtain resources, as well as shaping the ability of individuals or groups to act upon them. These factors include social class, place of residence (tenure now plays as important a part as locality), age, gender, ethnic group, disability and personal and collective values (Parry *et al.*, 1992).

Within this framework, consider the position of older people. For all older people there is the burden of ageism, whereby older people confront the negative attitudes of others and frequently come to internalise them until they become part of the accoutrements of old age itself (Gilroy, 1993). In addition, those who are poor and/or disabled and/or female bear multiple disadvantage, carrying, as it were, the conjunction of many forms of socialisation (Lonsdale, 1990). For many older people, age may simply exacerbate the divisions experienced in earlier life.

Planners and other professionals need to be aware of the ways in which all these groups have been marginalised and disempowered. They need to understand that power not only constructs a framework for dialogue but also defines what counts as knowledge and therefore what constitutes reality. Sandercock and Forsyth (1990), discussing women and planning, argue that power structures and the traditional epistemologies create inequalities by excluding certain groups as agents of knowledge and ignoring their ways of knowing. The importance of storytelling, listening, intuitive knowledge and symbolism in opening dialogue between communities and professionals has been acknowledged by feminist writers and radical planners. This acknowledgment of different ways of knowing and of multiple realities needs to be developed into pluralist, participatory forms which embrace the views and aspirations of older men and women. More fine-grained work is needed, examining the way older people articulate their concerns, if planners are to develop best-practice participatory forms. Some of the current attempts have not considered issues of power and the resultant tokenism is obvious to older people themselves:

> It is becoming a practice for local health authorities to hold so called forums as a substitute for consultation. They invite some older people to attend as a token gesture. We are usually outnumbered – and often drowned out – by non-elderly professionals who, however well-meaning, speak on our behalf. It is rare that in depth discussion can take place because of the nature of the forum and very often no report is produced or recommendation acted upon.
> (Sowerby, 1994, pp. 8–9)

If, as Leather (1997) asserts, the next group of retired people will be largely more affluent, they may well have more determination to use this affluence to give themselves greater control over their lives. They will want to be involved in decision-making and will be far less willing than previous groups to have decisions made for them by professionals or to have things done to them. It is therefore clear that we need to set up effective mechanisms now and examine the good models tried elsewhere.

One of the most effective was used in Vancouver, which based its highly acclaimed participative model for CityPlan partly on a project with older people carried out in 1991 (see Box 5.4). This project, entitled 'Ready or Not', aimed to plan for the considerable impact (challenges, demands and opportunities) that the ageing population might make upon city residents and services, to help prepare the city's administration for the attitudinal programme and structural changes required to deal with public involvement in planning differently from past methods, through using a partnership approach. Examples from outside planning show, on a smaller scale, what a little imagination can do to bring in the voice of older people (see Box 5.5). None of these need cost a lot in terms of resources.

What these strategies do cost is a change of attitude. We need to challenge our own ageism and find out from older people what this period of life is like. Without the deliberate creation of opportunities for older people to speak, we are unlikely to hear from them. This is not because older people have no views, nor is it because inter-personal communication presents too many difficulties, but because at present we make it hard for these people to be heard. For the future, older people may be banging on the door demanding a voice but we need to set in motion mechanisms of consultation and participation, so that as *we* plan to grow old we can rely on having a say, being listened to and our experience being valued.

Box 5.4 Ageing in Vancouver

Following discussions with city residents on how citizens wanted to work with professionals, workshops were set up in 24 neighbourhoods and more than 1,000 residents took part. In these sessions small-group work was used to facilitate discussion of a range of issues related to the impact of ageing and the groups decided on 1 or 2 priority issues for in-depth discussion. Residents were asked to identify ways in which the neighbourhood already met the needs of people as they age; what could be done to address anticipated needs; what could be done immediately to meet identified need either by themselves or with other neighbourhoods or with the city government. The workshops and the neighbourhood fora which followed led to a dual benefit. Firstly a Strategic Plan for Ageing was set out which had taken account of community priorities in arts/culture; community empowerment; education; environment/neighbourhood planning; health; housing; recreation; safety and security; and transportation. Secondly the process of empowering neighbourhoods and older people within them led to many workshop groups continuing to meet and devise ways of meeting the objectives they had outlined earlier. Many of these groups were awarded financial help from the Vancouver Community Building Fund to a maximum of $CAN 2,800 [about £1,400) to enable them to cover expenses for projects that would in some way contribute to improve social service provision or community development.

(McNeill, 1993)

Box 5.5 Alternative ageing-support strategies

A users' forum set up by a branch of Age Concern has as its members active retired people interested in improving local services. The forum meets monthly and takes up issues from a range of public services, including public transport, but individual members also take part in joint planning groups with statutory agencies.

Healthlink is a postal network developed by a Community Health Council in an inner London borough in order to reach people in the area who had difficulty leaving their homes both to contribute to making changes and also to find out more about local services. Members are kept in touch through regular mailings which ask for their view and experiences on a particular topic, and now has managed to get funding for a full-time worker to develop the organisation for the next three years.

The third project, a telephone discussion group, was set up by the research team themselves. Five elderly people living alone in rural areas 'met' for an hour each week for six weeks, discussing their experiences of current services and how things could be improved for them. At the fourth meeting, a manager for older people's services from the local social services department also took part, and went on to put the group's views at meetings with health, housing and voluntary sector services providers.

(Rickford, 1993, pp. 21–3)

6 Planning for disability

Barrier-free living

Linda Davies

This chapter discusses the unresolved relationship between 'town planning' and 'disability'. It discusses definitions of disability, reviews legislation and regulation, and provides a series of case studies. (A full account of the legislation, especially the 1995 Disability Discrimination Act is available in L. Davies, 1999; see also L. Davies, 1992, 1996; BBC, 1996; LGIU, 1995). A key theme is the move away from policies based on physical control (through the Building Regulations) towards those encouraging social access (through town planning policy).

The disabled

It is estimated there are over 6 million disabled people in Britain (OPCS, 1989; ONS, 1996, p. 151, Table 8.16: 'People registered with local authorities as having disabilities, by age') with a wide variety of forms of impairment and disablement, with those in wheelchairs constituting only 3 per cent of the total (as discussed in more detail L. Davies, 1999; Greed and Roberts, 1998, pp. 229–37) (see Figure 6.1). There are over three and a half million disabled women compared with about two and a half million disabled men. A related issue is the fact that the population is ageing; women live longer than men (cf. chapter 5; Martin *et al.*, 1988; Age Concern, 1993). By the year 2041, the ONS anticipate that nearly a quarter of the population will be over the age of 65 and 9 per cent will be over 80. Most of these elderly people will live out their lives in their own homes, and many will be affected by some impairment or impediment. This has implications for the design, type, and location of new housing in the future. Department of the Environment housing projections show an increase of just under 30 per cent in retirement age households in England in the twenty years to 2016 (DETR, 1998a).

Since the disabled and elderly represent a large sector of the population, covering all classes of people (a minority bigger than the entire population of some other European countries) it is no longer appropriate to adopt a segregated or divisive approach in which 'planning for the disabled' has often meant devising a special system for them, resulting, for example, in a separate external ramp being provided adjacent to a flight of stairs. What disabled groups have been pressing for, and the various professions in the built environment need to recognise and

Figure 6.1 Disability symbols. Although only a minority of disabled people are confined to wheelchairs, the wheelchair access symbol has come to represent them all. The visually impaired, for example, may not even see these symbols.

address, is the need for inclusive solutions in the built environment, which integrate disabled people as regards, for example, design considerations. Overall, this would produce a better built-environment, not just for disabled people but for all. Indeed, the Royal Town Planning Institute has recognised that the divide between disabled and abled is not clear cut, and that anyone can be disabled temporally:

> Disability includes a wide range of conditions: it covers more than the obvious such as blindness or confinement to a wheelchair. Breathlessness, pain, the need to walk with a stick, difficulty in gripping because of paralysis or arthritis, lack of physical co-ordination, partial sight, deafness and pregnancy can all affect a person's mobility in the environment. Access for the disabled will also benefit parents with buggies and the elderly.
>
> (RTPI, 1985, p. 1, para. 2)

Political attitudes are also changing. At the 1997 Labour Party Conference, Tony Blair said on 1 October: 'Our new society that we want to create will have the same values as it ever did, fighting poverty and unemployment, securing justice and opportunity . . . it should be a compassionate society, it must be a compassionate society. But it is compassion with a hard edge.' If the new Labour government maintains its commitment to compassion with a hard edge there is every prospect of success, both in terms of improvements to the planning process to achieve more equitable consideration of all the interests at stake in the use of land, and in terms of improvements to the product of planning (see Lock, 1998, p. 16). This augurs well for disabled people.

Disability

There are at least three models of disability: the medical, the charitable and the social (Swain *et al.*, 1993). In summary, the medical model is one in which disability is seen as an illness or as a permanent disability, set within the confines of the hospital or institution, and therefore unlikely to impinge upon the community or the built environment outside. Following this model a person might say, 'I cannot go into the museum or the cinema because my disability prevents me from climbing the stairs.' In contrast, recent 'Care in the Community' policies have precipitated the need to make the environment more accessible for a wider range of disabled people who have moved back into the community. Interestingly, the World Health Organisation (WHO) definition refers specifically to impairment, disability and handicap and concentrates on the individual's personal condition (Imrie, 1996; Hartrop, 1998) although, as implied in Chapter 10 the WHO is nowadays developing a more inclusive, sustainable approach to disability policy.

Second, the charity model of disability leaves little role for state intervention. The image is one associated with pity, embarrassment, do-gooding, dependency, sympathy and rattling collection tins. (But, the editor notes, with decreased state provision and privatisation 'flag days' remain an important component of provision.)

This contrasts sharply with the third model, the social model of disability, which is based on seeing people with disabilities as having human rights. This model draws on American civil rights disability movements, and is inspired by the Americans with Disabilities Act of 1988 (Manley, 1998). The emphasis is shifted towards seeing society's attitudes and thus the design of the built environment, as themselves disabling. According to this model a person might say, 'I cannot go to the museum or the cinema because the steps prevent me entering the building.' Thus, those working in the built environment professions have a major role to play. Disability groups have actively argued that all people should be able to gain access to buildings, with no fuss, no assistance, 'as normal', as (*inter alia*) workers, shoppers, theatre-goers and students. Ramps are viewed as cheap second best add on solutions, not something that disabled people should be grateful for (Imrie, 1996).

By the beginning of the 1980s wider political and theoretical discussion was taking place with regard to social models of disability, which were, in turn, challenging social theory (Oliver, 1996). It has been argued, for example that the disabled should be seen as a disadvantaged or oppressed minority group, whose unequal economic and social positions stem from discrimination and lack of access to power (Barnes, 1991). This model was enthusiastically received by many disabled people because it made an immediate connection to many of their own experiences, but not by all (see French, 'Disability, impairment or something in between?' p. 19, in Swain *et al.*, 1993). Its shortcomings have been discussed by Crow (1992). Finkelstein (1993) has stated that it may partially but not fully explain the social oppression felt by disabled people. Further terms of reference for this model are explained in Lonsdale (1990), Imrie (1996), Frechette (1996) and Bagilhole (1997).

Legislative powers

Whilst political theory might provide the motivation and justification for change, legislative change is the key in 'redesigning the built environment' because, arguably, of all the 'minority issues' discussed in this book 'disability' requires the greatest physical change to our towns and cities. A summary of legislation relating to disability follows (see L. Davies, 1999 for a detailed account). Significantly, much of the legislation is outside the realms of town planning, resulting in divided powers, so that many a disability-related planning issue is met with the reply, 'It's *ultra vires*, disability is not a land use matter.'

The Chronically Sick and Disabled Persons Act, 1970 (amended in 1976), first 'allowed' equal access for disabled people to *new* public buildings and places where education and employment took place. The Act also emphasised community care by strengthening the duties of local authorities to adapt homes. Section 4 first required builders to provide disabled access to and within buildings, with particular reference to parking and sanitary conveniences. This was, however, only required 'as far as is practical and reasonable': consequently, inaccessible buildings continued to be built and there was limited enforcement of the Act.

The Disabled Persons Act, 1981, required local planning authorities to draw the attention of developers to the relevant provisions of the Chronically Sick and Disabled Persons Act, 1970, and also to relevant design guidance such as Building Standards (BS) 5810, 1979, when granting planning permission (L. Davies, 1996). Limited central government policy support statements (DoE, 1985, which is DCPN 16) encouraged planners to negotiate with applicants to provide access and facilities such as car parking spaces for disabled people. Local authorities were also encouraged to designate a member of staff as an access officer, to provide a clear point of contact on questions of access for disabled people. In Scotland, all building control authorities (but not planning authorities) designated such an officer to ensure disability considerations were taken into account. The Local Government (Access to Information) Act, 1986, required councils to give 'the public' the right to attend and hear council and council committee meetings and read reports. Disabled people's special needs were recognised but not defined in the Act.

The next big milestone was the 1995 Disability Discrimination Act. At the time of writing, this Act is still coming into force in stages, but it has been much criticised as a compromise. Although it tightens some defintions and requirements, there are still 'get out' clauses, and the access standards required still only apply to new, not existing, buildings – that is, they are not retrospective. Indeed, the Act contains relatively little directly on the built environment per se, although Section 19(1) of the Act states (vaguely) that service providers should, amongst other things, remove or alter physical barriers. However, a series of policy statements have been made in various PPGs Planning Policy Guidance, produced by the DETR, previously the DoE, especially on general principles, such as those in PPG1 (DoE, 1997a) in which access for the disabled and for 'those with young children' are significantly linked; and on access to historic buildings (DoE 1997b, para. 3.28) and new houses (DETR, 1998a). As yet, though, there has been no PPG specifically on disability. This lack of emphasis is in contrast to more recent design guidance, such as the work of Julie Fleck, access officer in the City of

London, in producing a series of ground-breaking disability design guides for one of the most historic areas of Britain (Fleck, 1996).

Various more progressive local authorities have included disability-related policy statements in their Structure Plan and unitary Development Plan statements in respect of housing, employment and transport policy, rather than just providing a couple of ramps. Examples include the London boroughs of Islington and Waltham Forest, and Wakefield Metropolitan District Council (RTPI, 1993). However, generally, DETR guidance advises planners to look to the role of the Building Regulations as a means of providing disabled access (DoE, 1995b). Nevertheless, the Access Committee for England (ACE, 1994), and other bodies, such as the Centre for Accessible Environments, have continued to lobby both at national level, in respect of individual proposals in local areas, and regarding current issues such as access to historic buildings (CAE,

Table 6.1 Differences between the planning and the building control process

The planning control process	*The building control process*
Local policies and planning variations (this may result in varying provision and uncertainty for users)	National objectives and standards (people know what to expect)
Standards measured against approved plan	Measured against national criteria
Applies to most types and ages of land use and development	Only applies to new build, rebuild and major extensions, various exemptions
Developers/builders need to find out planning requirements	Developers know what is expected as nationwide
Long approval process before work starts	Once plans are deposited work can start or start with simple provision of notice
Work in progress seldom inspected	Work in progress inspected
Public can inspect plans and plan register	Plans are not open to public
Clients and public involved and think they know about planning	Decisions seen as technical and unlikely to be understood
Consultation and public participation	No such outside liaison
Councillors must approve decisions	Officers make final decision
Planners must consult with many groups	Only consult with fire service
Concerned with physical, social, economic and environmental factors and other on and off-site issues	Structural factors, fire and safety
Mainly land-use control and external control.	Mainly internal and structural design control.
Must advertise major changes	Can make changes, relaxations
Can approve phases of plan	Must approve whole scheme
Some control over future provision/ fate of conditions	Cannot control future management or maintenance of access features

1998; Palfreyman, 1998). Indeed, many local authorities liaise with local access and disability groups, thus increasing participation and inclusion within the planning system.

Building Regulations

The main source of regulation remains the Building Regulations, which put into operation the requirements of linked British Standards documents. In 1987, Part M of the Building Regulations, based on BS5810 on disability standards, was first introduced. It was extended in 1992 (RTPI, 1991b; DoE, 1995). Central government still looks to the Building Regulations, not the planning system, to impose requirements for implementing access for disabled people, as has been proved at appeal and at public inquiries 'testing' Development Plan policy statements (L. Davies, 1996). Thus, the social and physical dimensions of social town planning are pushed further apart. Not only do these regulations only apply to new or substantially rebuilt public buildings, they exclude domestic dwellings and many small-scale building alterations. Recent plans to extend Part M to aspects of domestic dwellings, including provision of ground-floor-entrance toilets and level entry to the main entrance of the dwelling, have met with opposition from some national bulk housebuilding organisations (HBF, 1995) but are seen as a sign of governmental commitment to implementing the duties of the Disability Discrimination Act, and thus to social inclusion and 'lifetime housing' (Daunt, 1991; Lonsdale, 1990). But the Building Regulation system itself is still considered to be flawed, not least because of the minimal wider policy content, the lack of public participation, its male domination and lack of integration with the planning system (see Table 6.1).

Case studies

This part of the chapter focuses on five case studies that serve to illustrate how access for disability has been successfully sought, and implemented, in the planning process. Importantly, these examples involved discussion and negotiation with disability groups throughout the process. It was purposely decided to include this somewhat 'nitty gritty' technical/physical section in a sociological book because, when one is discussing the social/physical divide in respect of disability issues, clearly 'space matters'. Indeed, the case studies demonstrate the interaction of the social and the physical in the design process. They also show the interplay of time, people, bureaucracy, law, tedium, culture and serendipity in the planning process.

Case study 1: Swindon Borough Council, local planning authority

The NatWest Bank, Swindon, wished to alter their frontage to a main pedestrian precinct by introducing a new ATM cash till and repositioning the entrance door (see Figure 6.2). The local authority, the then Thamesdown Borough Council,

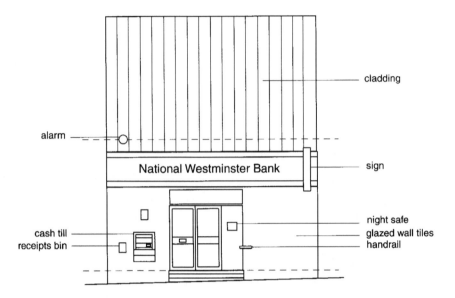

Figure 6.2 Former NatWest Bank frontage, Regent Street, Swindon.

now Swindon Borough Council, had in force a local policy, incorporated in the then consultative draft of the local plan. The policy required that the altered shop front had to incorporate improved access for all, including wheelchair users. One traditional method of doing this would have been to incorporate a ramped access to overcome the three steps up to the front door, but the constraints of the site prevented this. The frontage was on a gently sloping site and overlooked a busy pedestrianised street in the centre of town. There was physically no space for a ramp. Alternative solutions were sought by the local access officer, in consultation with the local disability group and the applicants. Some solutions, such as using the rear of the premises solely for access by disabled people, using a back door in a service yard adjacent to the deposited rubbish, were rejected outright by the access officer. Nevertheless, the bank directors were determined to follow a scheme with minimum alteration and cost, and persisted to the point of submitting their application requesting planning permission, without any provision at all for access for disabled people. This was refused by the planning committee, following their site visit. Several months later, the applicants returned and presented the planning committee with their revised shop front alterations, incorporating the recommendations of the access officer. The revised plans (see Figure 6.3) showed a reduction in the internal floor level, lowered internal cash dispensers and wide entrance doors with a flush access, all of which allowed equality of access to the bank for people with a mobility problem. The alterations are now in place and represent a small success for access for all to barrier-free living. They typically demonstrate the need for persistent negotiation and discussion as well as

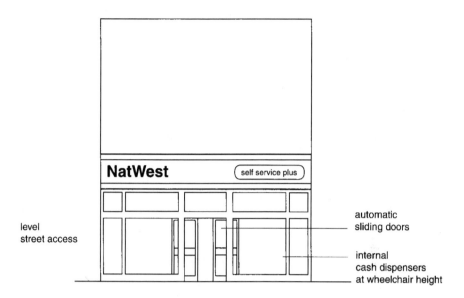

Figure 6.3 New NatWest Bank frontage, Regent Street, Swindon.

adherence to – often minimum – policy standards by the access officer on behalf of the local authority and the access group that he or she represents, even when challenged by large national organisations.

Another example of access achieved after lengthy negotiations, in the same town, can be found in the case of the Top Rank bingo hall, where the local authority was again involved. Many years earlier the cinema had been converted to a bingo hall; during its recent refurbishment the owners refused to implement access for disabled people. In the course of alterations to the entrance foyer, they proposed to introduce two steps inside the front door, in an area that had previously had a level access, and to alter the entrance door, all of which required planning permission. Negotiations with the local authority were protracted, largely because of disagreements. Finally, Top Rank constructed the two internal steps in the entrance foyer, without the required planning consent. The local authority began enforcement proceedings to remove the offending alterations. As time moved on, the Disability Discrimination Act (DDA) of 1995 came into force, thereby requiring Top Rank to comply with the legislation at some time within the following ten years. Reluctantly, the owners restored the flush access to the satisfaction of the local authority. Bingo still takes place, with unrestricted access for all.

This second example also shows the constant difficulty of persuading owners and developers of the need to consider access for disabled people in their designs, even in altering existing buildings. In this particular case, work that had previously been unfeasible for the owners to carry out for the local authority had become essential by December 1996, in order to comply with new central government legislation, namely the DDA, 1995, which came into force at that time. Although there is a phased time allowance for compliance with the legislative requirements, the promoters of the alterations to the bingo hall decided that they would carry out the work sooner rather than later. The effect of the Act was to lever reluctant owner/developers into carrying out work that they had no wish to do, even if it allowed access for disabled people. It is intended that the Act will nudge both alterations to buildings and the construction of new buildings towards barrier-free access for all, encouraging the development industry to do things that in the past were considered not feasible, or too costly (Swindon Borough Council, 1996, 1998; Swindon Access Action Group, 1998; Thamesdown Borough Council, 1995a, 1995b).

Case study 2: Swindon Railway Heritage Museum Advisory Group, the Swindon Access Action Group

This is an example of positive and early consultation with a local access group on a community issue, in the railway museum development in Swindon (Swindon Borough Council, 1997/8). The editor notes that this example also shows how 'disability' can be linked to planning for tourism and cultural activities. The Swindon Access Action Group aims to promote effective consultation and build working relationships that will help to achieve an accessible environment for all. To this end, the Group wants Swindon to be a fully accessible city for the twenty-first century, stating that it will oppose any proposals that restrict access and the integration of people with disabilities. It also expects to be a consultee in all planning and design matters, and at an early stage. The opportunity to put all this into practice arrived with the proposals for the Railway Heritage Museum, largely financed by the allocation of around £11 million from the Heritage Lottery Fund in recognition of Swindon's worldwide reputation as the home of Brunel's Great Western Railway Works. The intended opening date of the museum is the year 2000, and work is under way.

The Museum Action Group comprises members of the Swindon Action Group, the Hard of Hearing Club and a representative of the Board of Directors of Museums, led by two council officers, one representing the Railway Heritage Project. The other, with overall responsibility, is the local access officer. There are some twenty members in all. Opportunity has been afforded by Swindon Borough Council, the local authority, and the contracted design team, for the Action Group to become involved at the early stages in the planning and design matters of the renovation and reconstruction of the building, which has a Grade 2 listing, as being of architectural and historic interest.

Access to the new Swindon Borough Council Offices, where the regular meetings about the museum take place, is simplified by the fact that the newly built offices, opened in 1997, are constructed in accordance with the latest access design requirements. Parking for disabled people is adjacent to a ramped access; the entrance door is extra wide and is automated by a push button for wheelchair users.

At the regular monthly Action Group meetings, chaired by the local access officer, the following devices are available within the room to facilitate the full participation of those with a disability: a lip-reader, a Braille agenda, touch reading, a three dimensional model of the proposed heritage museum for a blind representative who is helped by explanations from an assistant, an infra-red induction-assisted hearing loop and a centrally located transmitter with neck loops and handsets.

Once familiar with the proposals for the Heritage Centre, the Action Group is able to contribute to the debate. Suggestions which have emerged from the meeting regarding the design and operation of the Heritage Centre are as follows:

- Ticket desk to provide a variety of ways of presenting entry prices and other information, rather than written lists only. The availability of, for example, a counter loop for the hard of hearing, a 'talking price list' and staff at the desk to deal with customers' extra needs will all aid access for disabled as well as able-bodied people.
- Staff to receive disability awareness training, in order to be prepared to help customers who have access problems.
- Lift access to all floors of the building.
- Lift access to the third floor of the building, where the school lunch room is located, to help disabled schoolchildren reach this facility.
- Lift contractors to be available at all museum opening times, in case of breakdown of lift.
- Lift design to incorporate sufficiently wide doors, manoeuvring space, 'invisible eyes' and adequate length of opening times for wheelchair users.
- Revolving doors to be avoided.
- Hands-on computer technology room to have large-scale screens and to be generally accessible by the physically impaired.
- A speech processor to be installed for those with a speech impediment.
- Handrails to be fitted in corridors and in large exhibit rooms.
- Foundry room with visual and sensory experiences, may be unsuitable for disabled people: an escape passage is therefore needed.
- Height of written displays to be at comfortable level for wheelchair users.
- Scissor lifts to help change levels, rather than ramps (slope and length too demanding on internal space).
- All corridors and doorways to be to recommended standards, in accordance with the Building Regulations.
- How disabled people will get on and off the moving turntable, a museum feature, to be investigated. A stop button is suggested.
- Need for many touch exhibits.

In this way, by having regular discussion meetings with a local disability group, the access officer can act as the interface between their needs in a new building in the town and the designers and other colleagues on the council. As a result of the close working relationship between the users and the professionals, the views of the users are considered and attempts are made incrementally, throughout the early stage of the design process, to accommodate essential design criteria which may only be apparent to this group of users of the proposed building. This inclusive approach taken by the designers and developers means that the resultant design should meet the needs of the users, and not require expensive add on features at a later stage, as so often happens.

Case study 3: City of London Access Group

Another example where personal access and mobility is important is in the realms of transport planning. (The editor notes that this issue links to Geoff Vigar's chapter, on transport for people.) In the 1990s over 10 per cent of the people living in Britain had some degree of disability which made it difficult for them to travel by bus. Population projections show that an increasing percentage of women will outlive men. The fact that fewer women than men hold a driving licence (53 per cent as opposed to 84 per cent (Labour Party Manifesto, 1997) makes it evident that by the year 2000 older women will rely heavily on public transport to meet their mobility needs. Personal mobility not only affects people with disability; it is also a gender issue and therefore requires, in addition, to be considered as an equality issue.

It is estimated from previous research by the author that about 20 per cent of bus passengers are over retirement age (18.2 per cent of the population are over pensionable age; see DoT, 1994 and in off-peak periods the majority of public transport users have some form of impaired mobility, not just through disability or age, but as a result of being encumbered with shopping bags, trolleys or heavy luggage, or because they are travelling with small children and a pushchair. Retaining such passengers and attracting more of them, and more passengers in general, to public transport, provides an important opportunity to increase bus use and hence revenue, as suggested by the Disabled Persons Transport Advisory Committee.

Because of the importance of mobility by public transport, the City of London Access Group has regular discussions with the London Transport Disability Unit, who also attend its meetings. At one such meeting (26 November 1997) the low-floor bus project and the London Underground were discussed (City of London Access Group, 1997/8; City of Westminster, 1997; Fleck, 1992).

It is widely accepted that many disabled and elderly people find traditional designs of public transport difficult or even impossible to use. Additionally, there are other features of transport systems which create barriers to disabled and elderly people. With this in mind the low-floor bus project commenced on five routes in London in 1993. The buses have step-free doorways and the addition of a powered ramp at the second door, making them accessible for wheelchair users. In order for the ramps on the buses to be effective the bus needs to be parked directly alongside the kerb. In central London this is often made difficult by illegally parked vehicles or because the driver cannot easily park close enough to the kerb. To alleviate these problems the installation at bus stops of cameras and Kassel Kerbs (raised kerbs pioneered in Kassel, Germany to facilitate tram access for disabled passengers) is being considered. As is often the case in the provision of facilities for disabled people, the existence of low-floor buses benefits all passengers, especially those pushing pushchairs and prams, the otherwise ambulant elderly, who may simply find it difficult to lift their legs, and toddlers. As well as providing access for disabled people, the absence of steps on low-floor buses makes them more attractive to many thousands of passengers whose mobility is impaired. They also attract more passengers and increase the use made of public transport by, for example, parents travelling with small children or people with heavy luggage, for whom the low floor makes bus travel a civilised alternative to the private car, taxi or minicab. (DoT, 1994). Previous national research on the drafting of unitary development plan policies on access for the disabled showed that the most popular policy area was in the field of transport and movement (85 per cent), more so than housing (80 per cent) and shopping (76 per cent). (Davies, 1996). Davies' research showed that the more popular (draft) transport policies in the unitary plans included access and movement within towns, access to public transport and adequate car parking facilities.

As the City of London Access Group was quick to point out, another of the benefits for disabled people was that the provision of an accessible bus allowed them the opportunity for spontaneous travel. Most other forms of transport that are provided with disabled people in mind, such as dial-a-ride taxis, require there to be some planning ahead on the part of the disabled person. The provision of an accessible service bus means that they, like everyone else, have the opportunity to be spontaneous in their movements.

London Transport will, however, require some ten years to put into operation a totally accessible low-floor bus network throughout London, so freedom and spontaneity to access buses and travel is not yet universally available. In the meantime, experiments continue with other types of accessible transport, such as the introduction of the first double-decker low-floor bus, which was launched in central London in 1998. The existing mobility buses will not be withdrawn from service until the entire area currently served by them has been replaced by a network of low-floor buses.

Another mobility concern of the City of London Access Group is the perennial problem of travel on the London Underground. The London Underground is one of the oldest and deepest underground systems in the world, if not the oldest and deepest, which makes it very difficult to make accessible to people with disabilities. The problem is further complicated by the fact that although some stations have provision for access by wheelchair users, others do not. This prevents or inhibits people with disabilities using the Underground, as they cannot always use it between the stations of their choice in the way that an able-bodied person would do. Future investment by the new Labour government, irrespective of the funding arrangement, will need to give consideration to the needs of disabled people.

According to the minutes for 26 November 1997 (City of London Access Group, 1997/8) the City of London Access Group was reminded that London Transport had made some provision for the mobility of people with disabilities on the Jubilee Line Extension, which was designed following the same format as the Docklands Light Railway, with lift access at all stations and minimal gaps both horizontally and vertically between the platform and train.

The Access Group took the opportunity at the meeting, in much the same way as the Swindon Access Group, to make some suggestions for a number of possible areas for improvement to the existing public transport in central London. These were as follows:

- The signage indicating priority seating for elderly and disabled people on public transport should be in larger print, to increase the legibility of the sign and give it more prominence, and it should be repositioned higher in order to remain visible when the seats were occupied.
- Travel on escalators for disabled people should be improved by extending the handrails. (The group was pleased to hear that the new Jubilee Line escalators would have a lead-in of eight steps prior to the descent).
- There was a need to vary the length of time Underground doors remained open; Underground drivers should have the opportunity to alter the length of time that doors remained open or closed, especially for the elderly who might be using canes or walking sticks. (The Access Group was informed that the London Underground was conducting experiments with platform-edge doors, as are often used on the continent. It was confirmed that Underground drivers did have the opportunity to vary opening and closing times of doors and should ensure that everyone, not just the elderly and disabled, had safely boarded the Underground).

A significant number of low-floor buses were operating throughout the United Kingdom by the end of 1996. Initially, most were 12-metre single-decker buses, but low-floor minibuses are now proving to be very popular and manufacturers are turning their attention to this market. Another area of the built environment that is gradually being adapted, in part to meet the needs of disabled people, is in the area of access to shopping, known as Shopmobility.

Case study 4: Shopmobility, Bristol City Council

Shopmobility schemes provide powered scooters and wheelchairs, and manual wheelchairs, to enable people with limited mobility to shop and use other facilities in the relevant shopping area. An escort service is often available for people with visual impairment and for wheelchair users. Anyone with limited mobility can use Shopmobility, whether through permanent or temporary disability, illness, accident or age; it therefore includes a wide range of people. Local authorities throughout the country are increasingly proactive in introducing Shopmobility schemes in their town centres. The schemes at Broadmead City Centre shopping area in Bristol and in the Royal Borough of Windsor and Maidenhead are typical.

The Bristol scheme operates for part of the week, from Wednesday to Saturday, and arrangements can be made for Thursday late-night shopping. Training in the use of wheelchairs and powered scooters is given and for insurance purposes applicants must register with the Shopmobility Centre. To allow users to find their way around the Broadmead Centre, a city centre Access Map is available showing pedestrian areas, the Blue Badge and the Orange Badge national car-parking scheme for disabled people, 'dropping off' car parks, disabled people's toilets, safe places to cross the road and accessible pedestrian routes. As Bristol Shopmobility is dependent on charitable support, it does make a small charge for the service and monies raised through hire charges are reinvested to purchase new mobility aids.

In Bristol, local authorities have entered into a partnership, 'working together for access', with local private sector and voluntary organisations to make Shopmobility happen. Staff and revenue funding to support the venture are continually difficult to find but recently Bristol Shopmobility acquired charitable status. In Maidenhead, by contrast, access improvements to its central shopping precinct have been less directed by the local authority and mostly achieved by the action of the local access group, who have negotiated independently with the business community about improving their facilities.

It is important, however, that the planning system physically supports and co-ordinates the introduction of the growing number of Shopmobility schemes throughout the country, by, for example, the introduction of planning policies or the successive granting of individual planning permissions requiring pedestrianisation, flush access and shop entrances of adequate width. In this way incremental improvements to the built environment will take place, from which all can benefit.

Conclusion

This chapter has considered a wide range of issues concerning disabled people and their need for equality of access to the built environment. Barrier-free living should be a right, a civil right, for all. Throughout the early and mid-1990s, the opinion of agencies working on behalf of disabled people was that central government had not been sufficiently bold in providing statutory support for the disabled in the built environment. The 1998 announcements that Part M of the Building Regulations would be extended to incorporate new housing, and require that it be designed with disabled people and their mobility in mind, has been heralded as a landmark success for disabled groups throughout Britain. Previous criticism of Part M of the Building Regulations, 1992, regarding the fact that Part M appeared to be geared to the needs of property developers and intended to minimise their costs (Imrie, 1996) appears no longer to apply.

Although the Discrimination Disability Act of 1995, represented an important landmark for disabled people in terms of access to goods, facilities and services, it contains large omissions. The British Council of Organisations of Disabled People (BCODP, 1997), in response to a speech made at the Labour Party Conference in 1997 by Andrew Smith, MP, who was to become Minister for Employment and Disability Rights in the Labour government, demanded the repeal of the oppressive Act. BCODP considers the Act an insult to disabled people, largely because it is discriminatory legislation set within a medical, not a social, definition of disability. It looks to the social definition to challenge what it regards as the specific forms of oppression experienced by disabled women and men in all Britain's communities. Such legislation must lead to the planned removal of disabling barriers and guarantee full civil rights for all disabled people (BCODP, 1997). A strong welcome is given by the BCODP to the setting-up of a ministerial task force to write a new Bill for comprehensive anti-discriminative legislation, provided that the task force is properly representative of, and accountable to, organisations for disabled people.

There is very little evidence of strategic-level guidance. Counties, districts and unitary authorities provide access for disabled people in a very piecemeal and reactive manner. This is partly due to the fact that local authorities' ability to contribute to providing a service to disability groups is restricted by central government budgetary controls. In these circumstances partnership schemes emerge in which the local authority is a participating member, such as the Windsor and Maidenhead Access Group, the Worthing Access and Mobility Group, Key London Housing Unit, York Dial-a-Bus and York Wheels. These local groups are often formed as a response to pressures exerted from other groups and charities such as RNIB, Mencap and local equivalents, who work directly with private businesses to guide development and promote awareness of access issues. Advances in disabled people's rights have often originated from the actions of the disabled people's movement.

Access officers have a valuable role to play in raising awareness among the business and other community sectors and in arranging inspections of premises to ensure that they comply with access criteria (Access Officers' Association, 1996). These criteria should be set down in supplementary design guidance and cross-referenced to the Development Plan. Access officers should lobby for change if standards are inadequate.

Perhaps the greatest argument for designing for a barrier-free environment stems from the fact that there is an ever-ageing population, many of whom will inevitably be affected by some form of impairment or impediment. As stated in the introduction, physical disability currently affects over 6 million people in the United Kingdom. Age Concern's forecast (Age Concern, 1997) shows that the increase in an ageing population will become more rapid in the first decade of the new millennium, with a rise of almost 1 million (9 per cent) by 2010. Many of these elderly people will suffer from disabilities.

The great housing debate of the late 1990s has been 'where shall we live?' An adjunct to that question should be 'how shall we live?' At a time when the quality of the built environment is under review should central government not now be requiring local authorities to provide grant aid to those who wish to bring their existing housing accommodation up to a fully accessible standard? A national campaign is required, headed by the ministerial task force, to provide for all our housing needs for the next millennium.

As suggested earlier, the issues of disability and its solution in the built environment ought to be tied in to the new government's holistic approach to dealing with cities and the redevelopment of the built environment in general. This would ensure that the approach to urban areas involved not just planning but was inclusive of all relevant professions, involving related disciplines concerned with education, planning out crime, housing renewal and design of the built environment in general. In order to be meaningful to all, town and country planning must understand and embrace the social as well as the physical needs of all those affected by it.

7 Transport for people

Accessibility, mobility and equity in transport planning

Geoff Vigar

This chapter looks at UK transport policy and its social implications. It discusses three main elements. First it looks at key mobility trends in the UK in recent years. Second, it analyses recent policy responses from central and local government. Third, and most significantly, it explores concepts of accessibility, mobility and the potential for principles of social exclusion/inclusion to put a social agenda back into transport planning. Recommendations are made as to how a social agenda might be mobilised in practice and what it requires of those in transport policy communities and other areas of government.

Mobility trends

Movement in the UK in recent decades has been characterised by increasing ownership and use of private cars and a decline in the distances travelled on public transport, by bicycle and on foot. In the last decade there has been no significant increase in the number of journeys undertaken by either goods or people, but average distances travelled are increasing and virtually all growth in distance travelled, as Table 7.1 illustrates, is by car, van and lorry. The average person travelled 27 per cent further in 1996 than in 1986.

It is anticipated that the growth in road traffic will increase unless policies can be implemented to curb these trends. The environmental consequences from this increase in road transport in the form of exhaust emissions, noise and visual intrusion, local air quality and the effects of traffic on neighbourhood cohesion has provoked considerable concern.

Past transport trends are in part a result of previous government policy relating to land-use planning, road construction and approaches to the public transport network. Such policy responses are discussed below. Thus the decline in distance travelled by public transport, for example, is in contrast to many other countries in Europe which, whilst experiencing growth in private car travel, have also witnessed growth in the miles travelled by public transport.

Amongst politicians, transport and land-use planners, and to a degree the general public, a consensus is emerging which considers that continued growth in road traffic cannot continue. Such growth has 'resulted in a failure to reconcile the movement needs of different user groups and their competing demands for

Table 7.1 Passenger kilometres travelled by mode in historical perspective

	Bus/coach	Car/van/taxi	Rail	All (includes motor cycles and bicycles)
1956	89	91	40	245
1966	67	252	35	369
1976	58	348	33	453
1986	47	465	37	566
1996	44	620	38	717

Source: DETR (1997a), Table 9.1.

movement space' (Hine, forthcoming). The implications of this competition for scarce movement space are discussed below.

The critical task for this chapter is to assess the social impact of this additional movement and its relation to land-use planning. Whilst transport planning is often considered to be the realm of engineers and economists, land-use planning has begun to play an increasing role in the consideration of transport policy. Land-use planning policy is vital in that the location of facilities and opportunities in space closely relates to people's opportunities to access them. Thus the increases in the length of trips mentioned above may be a result of a forced choice resulting from the closure of local facilities or they could equally be an expression of improved mobility, facilitated by improved transport services or people's willingness or ability to pay for such services. Similarly, a decline in the numbers and lengths of given trips may be a result of a decline in the affordability of such journeys, or it may be that remote accessing of a service via telecommunications is substituting for travel. It is impossible to interpret such trends at a macro-level. This promotes attention to micro-levels of analysis and people's day-to-day activities rather than transport planning's traditional focus on flows accumulated, trips and models.

The policy response

Barton (1998a) notes that the transport planner traditionally looks at transport problems from the point of view of mobility, rather than access. Given that the car has been more and more the mode of choice, transport planners' attempts to increase mobility have typically meant freeing roadspace from congestion to enable the faster movement of goods and people. This view was reflected in the working practices of transport planners for several decades during which the prime concern was to predict the amount of traffic that would want to travel along a given road in the future and draw up plans to cater for it, a so-called 'predict and provide' approach, (Owens, 1995). Whilst this paradigm has been criticised from the 1960s onwards, for a number of social and environmental reasons as well as for its (lack of) intellectual robustness it survived as the dominant *modus operandi* for transport planning into the 1990s (Vigar *et al.*, forthcoming, ch. 6).

There is evidence that the predict and provide paradigm has now largely been displaced (Goodwin *et al.*, 1991; P. Goodwin, 1997; Guy and Marvin, forthcoming; Vigar, 1996). Indeed, despite the dominance of this approach, the picture of transport planning was always more complicated than slavish conformity to 'predict and provide'. Concepts of demand management were present in the 1960s and have underpinned much work in towns and cities since then. However, it has taken some time for the notion of predict and provide to be usurped as the dominant paradigm. Owens (1995) suggests that the new emergent paradigm for transport planning might be termed 'predict and prevent', in that the focus is now on predicting future demand for road travel and looking at ways of preventing the demand from being realised, through demand-management measures. The emergence of this new paradigm appears to be provoked by concerns over the impact of future road traffic on the environment, the inability of even massive road building programmes to tackle such an increase in traffic, and the cost of road construction as a solution in a period of tight fiscal restraint.

National transport policy in the UK as a result shifted in the 1990s and is now aimed at promoting choice in transport through increasing the relative advantage of means of travel other than the car, especially walking, cycling, and public transport; at reducing dependence on the private car; and at increasing the competitiveness and attractiveness of urban centres against peripheral development (DoE, 1994a, paras 4.1–4.2). This view is given a new dimension through the provisions of the Road Traffic Reduction Act, 1997 and the Road Traffic Reduction (Targets) Bill, 1998. These two pieces of legislation require local authorities to put in place plans to reduce traffic levels in their areas, shifting attention from demand management specifically to demand reduction. However, the extent to which traffic levels will be reduced is open to question, given that commitment by local authorities will be variable, and both pieces of legislation stop short of either specifying targets or giving local authorities any powers or resources to implement them.

Many local authorities are, however, pursuing demand-management policies, principally through focusing on four elements:

- reducing the distances between people and facilities;
- reallocating existing roadspace to prioritise public transport, cycling and walking;
- better integration of public transport modes to achieve a seamless journey;
- appeals to end users (principally car drivers) to take a greater social and environmental responsibility.

(Banister, 1997)

Such a radical change of policy throughout local government (in a comparatively short space of time) will be contested in localities across the UK. Traffic will not easily be reduced as car use is deeply embedded in people's lifestyle choices and political commitment to such a project will vary between local authorities. Such a change goes against the views of many local politicians who perceive, with some justification, that unhindered car access is essential for local economic prosperity. Such a change also challenges the culture of transport planners in local government who have spent the last forty years focusing on mobility, principally through a free-flowing road network (Whitelegg, 1997; and see Pemberton and Vigar,

1998; Vigar *et al.*, forthcoming, ch. 6 for case studies). However, the 1998 Transport White Paper goes some way towards setting out how this might be achieved, going further in attempts to make car users pay the full cost of their transport decisions through taxation, potentially paving the way to give local authorities more powers with which to tackle these issues.

The legacy of past policies

The problem of local-level implementation is compounded, as the legacy of previous transport policies and socio-economic trends lives on in localities. The 'predict and provide' approach of the 1960s and 1970s led to attempts to provide for increasing car use in urban plans through the construction of roads in urban areas. These policies had a number of consequences. One was physically to dislocate inner-city communities, because of the barrier effects of new roads. Such roads often led to pedestrians having to travel greater distances, through underpasses and over walkways. Such roads also implied a physical danger to those not in cars and led to increases in local pollution levels. Many cities today, whilst not constructing physical barriers on the same scale, are still reluctant to restrict car travel and favour other modes to a great degree. This can be evident in the absence of pedestrian crossing facilities, the timings of such crossings, and a lack of attention to the needs of cyclists and bus users.

The planning interventions of the past have had a disproportionate effect throughout society. Other trends have affected the accessibility patterns of different societal groups. The last few decades have witnessed a decentralisation of employment, retail and leisure activity in urban areas, taking advantage of a relaxed land-use planning regime in the 1980s and a lack of congestion on the periphery of urban areas. Employment, retailing and leisure uses have at the same time become concentrated onto fewer, larger sites. Thus there has been a tendency for the distances travelled in order to access employment and services to increase throughout this time. Decentralisation has had the effect of undermining the possibilities for cycling and walking, and has made the provision of public transport to service new peripheral uses more and more difficult. Trends towards more flexible labour markets and double-income households further encourage the car as a tool for commuting. People change jobs more often than they change residential location and the result is often increased commuting and difficulty in bringing residential and employment locations together. These points highlight the need to focus on accessibility rather than mobility. This debate is explored below.

Equity and accessibility in transport planning

The devotion of policy attention to car-based travel is to the detriment of those without access to a car, with major social town planning implications. Those without regular access to a car are dependent on walking, cycling and public transport. As illustrated above, central government policy is now geared to promoting these modes, predominantly for environmental and financial reasons rather than social

ones. However, the legacy of past policies lives on and many inequitable conces-
sions remain, such as company-car tax advantages. A debate continues as to the
extent to which car taxes in general meet the true costs of car travel, when the
health implications, such as exhaust emissions and accidents, are considered.

Many people cannot fulfil their desires to access goods, services or opportunities to visit friends. People can be considered to be 'transport disadvantaged' or
'transport poor' when there is a 'discrepancy between their transport opportunities and their transport needs' (de Boer, 1986, p. 137). The question then arises
as to how we might assess need. Any assessment of need is essentially normative
but will involve an assessment of existing activity patterns and a consideration of
the space–time order of activities – the 'where' of things and the times taken to
access them for given individuals. A critical question then arises as to what services
are considered necessary. These will vary from individual to individual, but may
include educational facilities, churches, banks, health care, shops, and community
and leisure facilities. So assessing need is not an easy concept. What can be said,
is that if existing activity patterns are examined there are clear and widespread societal imbalances in the distances travelled and the services accessed.

The biases inherent in such processes are conditioned principally by income.
There is a direct correlation between income and the distances travelled per year.
The concept of equity is therefore critical in assessing the wider implications of
transport policy choices and subsequently those who are able to satisfy their
mobility needs and those whose ability to access opportunities is limited.

Whitelegg (1997) suggests there are three links between transport and equity:

- unequal distribution of finite natural resources between and within countries;
- unequal access to opportunities that favour the needs of certain groups;
- unequal distribution of the negative impacts of transport through society.

Each of these points is now considered in turn.

Global inequalities

On a global scale, the current transport trends of developed nations are inequitable,
not only in terms of the use of scarce finite resources such as oil but also in the
production of transport by-products such as exhaust emissions. Such issues provided
a major problem in negotiations over limiting the production of greenhouse gases at
the Rio Earth Summit (UNCED, 1992; Keating, 1993) and the United Nations
Convention on Climate Change held in Kyoto, Japan in November 1997
(UN, 1997; and see http://www/unfccc.de) as third-world nations objected to
certain developed countries failing to address their own increasing emissions.

Access and opportunity

The second of Whitelegg's points is perhaps the most pertinent in the context of
this discussion. It stems from transport rarely being an end in itself but, more

usually, a means to attain certain services or opportunities. A social planning agenda would seek to maximise people's access to opportunities. The absence of such opportunity is closely linked to the concept of social exclusion/inclusion. Given the current attention paid to issues of social exclusion at all levels of government, transport thus becomes a critical component.

So, certain groups will require particular attention. Access to car transport is strongly associated with income and gender (Root, 1996). Those on low incomes may be unable or unwilling to run private transport and thus are dependent on public transport, cycling, walking and lifts from car owners to access opportunities. Even accessing public transport may be an issue for some, such as the disabled and carers of children. Women, children, the elderly, and the less well-off tend to have lower levels of access to cars. They thus tend to be disproportionately affected by policies to provide for improved mobility for cars and by trends in land-use toward decentralisation and the dispersal of activities, as accessing such locations typically necessitates driving to them. Access to opportunities for such groups is as a result heavily conditioned by proximity to services and/or the availability of cheap public transport. The elderly are of particular concern in that 67 per cent of the over 70s do not have a driving licence (DETR, 1997a, Table 3.15). Even in households which own one car, the 'family car' is often used for the commute to work and other household members will not have access to it for much of the day, necessitating the use of other modes. In summary, accessing opportunities is less likely to be a problem if one has either a car or a relatively high level of income. It becomes a problem for those with low incomes and/or no access to a car. These issues are compounded in rural areas, where there is often little or no access to public transport. The ways in which poor accessibility can limit opportunity are discussed below.

Access to employment opportunities, especially for those in localities characterised by high unemployment and for the long-term unemployed, has long been a concern for policy-makers. The dislocation of employment and residential location is a continuing trend. Again, certain groups are particularly affected. Car ownership may be out of reach for those on low wages. In a study in rural Oxfordshire, 56 per cent of 16–29-year-olds said they had been prevented from applying for a job because of difficulties getting to it (Root, 1996). The location of employment opportunity nearby or within reach of low cost public transport is thus very important. Indeed, when public transport fares were reduced substantially in Sheffield and London in the 1980s the unemployed were one societal group that benefited considerably. They were able to widen not just the area in which they would apply for jobs but also the range of information they could access in seeking a job, as cheap fares meant they could travel more freely, thus widening their job-search patterns (Goodwin *et al.*, 1983). For other groups, such as those with family-care responsibilities, the key issue may be one of time, as the demands of employment will put pressure on care responsibilities and the high cost of help with such responsibilities. Again, proximity to employment opportunities will be a key consideration in taking employment under such circumstances, as this proximity will generally tend to reduce commuting time.

Whilst access to employment locations is a critical factor, access to services and leisure opportunities is a vital component of people's quality of life. As Barton (1998a) notes 'except for men in the 30–59 age bracket, the most important reason for travel (judged by distance) is not commuting but leisure activity', (p. 134). The closure of shops and banks in rural areas has long been an area of concern and this trend has extended to urban areas in recent years as shopping provision has concentrated in larger units, often away from residential areas, resulting in the closure of corner shops and small supermarkets in traditional shopping areas. This process can present real problems for those without access to cars. In the 1990s the continued withdrawal of financial services from rural communities and the trend towards the closure of branches in the secondary shopping districts of urban areas has provoked similar arguments. Under the Sheffield and London low-fares policies highlighted earlier, cheap public transport also meant that low-income households were able to exploit low-priced goods and services in parts of the city previously inaccessible because of the high cost of getting to them.

Trends in other public services are also a key element in discussions of accessibility. The trend for many years has been for health and education authorities looking for economies of scale to close smaller units of provision and to concentrate facilities at one site. This again raises questions of accessibility, especially for those without access to cars. This tendency is part of a wider drift towards managerialism in governance, in which different areas of government make decisions based not on what will cost the public sector more overall, but on what will cut that particular government department's costs. The division of government into individual cost heads is an understandable trend but it contributes to the perpetuation of social exclusion processes. Transport is just one of the many aspects that slip through this fragmented net of local governance in such circumstances. Such an approach privileges economic considerations over others, with the result that the Treasury has become 'the *de facto* arbiter of public priorities in domains formerly governed by less exclusively econometric articulations of the public interest, frequently exercised through administrative discretion' (Grove White, 1997, p. 27). This again tends to place pressure on certain sections of society, such as carers (who are likely to be women), who have to escort those for whom they have responsibility over longer distances to schools and hospitals, diminishing their own opportunities to engage in other activities. In a similar vein, the introduction of parental choice over schools has increased overall distances travelled to such facilities – longer distances tend to imply more travel by cars: as a consequence traffic heading for schools is now a major contributor to congestion in the morning rush hour in urban areas. Banister (1997) proposes a social audit as a way to assess the impact of these policies and get beyond the narrow, merely financial cost-driven assessment procedures that are currently used. Such an audit could be capable of assessing the economic costs transferred to the users of the service and other elements of the public sector, as well as assessing the social and environmental consequences of such decisions.

There are a number of other tools in the land-use planner's toolbox for tackling these issues. The development of accessibility standards for assessing planning applications for new development is a way of accommodating these concerns. Such standards require that new development be located only in places that fit a set of requirements in terms of access by alternatives to private cars. Similarly, guiding development to locations close to public transport nodes, as well as higher density development, will all help to achieve social and economic objectives. Many local authorities have embraced such issues. However, whilst this is welcome, new development only constitutes a small proportion of the built environment and there is a need to develop other measures in tandem with the application of such standards. Existing residents remain 'accessibility poor' and there is a danger, as planners rush to meet environmental objectives, that they do not consider adequately those residents in the existing housing stock.

Other solutions are purely transport related, such as the provision of subsidised public transport and funding the low-fares policies mentioned earlier. However, there are limits to what local government can now achieve in this regard. In finance terms it is heavily dependent on the spending priorities of central government. The franchising of socially necessary bus services is a local authority function but funds for such services have been diminishing, despite rural services being given a boost in the 1998 budget. Concessionary fare schemes are another means to subsidise travel for certain groups such as the elderly and the unemployed, but the extent of these schemes is determined by local authorities and many local authorities do not operate them at all. In addition, as public transport has been privatised there are few ways in which local government can promote public transport provision. Entering into partnerships with bus and rail companies is one of the few options open to local authorities in this regard, but their capacity to contribute is often limited.

Accessibility poverty

Whitelegg's third and final point referred to the unequal distribution of transport's negative impacts. Increasing mobility for some groups, such as car drivers, may damage the potential accessibility for other social groups. The construction of roads forms physical barriers to the movement of pedestrians and the use of cars damages air quality, makes pedestrian and living environments less safe and can hinder the movement of public transport. Clearly there is an equity question here, in that it is generally the poorer sections of society who do not have access to cars and it is they who are often damaged by provision for and use of cars. Again, children and those charged with looking after them are disproportionately affected by such activity, as children are more vulnerable to road traffic accidents and greater supervision is needed as a result of car use. Furthermore, many studies have indicated that interactions between neighbours as a whole are 'inversely related to [the] amount of traffic [in an area]' (Barton, 1998a, p. 136). Indeed, in many instances the construction of roads to facilitate economic development has damaged further the quality of life for residents in certain areas, accelerating the process of urban decentralisation and thus achieving the opposite of what was intended (Whitelegg, 1997).

So how do we approach the problems of 'accessibility poverty'? One of the central points in this book concerns spatial and aspatial policy divisions. In transport it is clear that some policies are aspatial. Improved access for the mobility impaired is in part aspatial, in that it should be a principal embodied in the working practices of transport planners, architects and all built environment professionals. But the principle of social exclusion shows the spatiality of notions such as 'quality of life'. The poor range of services available to certain sections of society living in particular places, and the difficulties many have in accessing them is an important consideration, and not just for transport planning. Such areas are characterised by distant services, little or no access to the use of a car, and poor or non-existent public transport services. Whilst for those with cars distant services and lack of public transport rarely present insurmountable problems, for those without a car or the ability to pay for alternatives, they can materially affect quality of life. A focus on quality of life and accessibility therefore provides the necessary hooks for such policy attention.

Perhaps particular attention needs to be placed here on low income rural residents. Whilst transport issues as a component of social exclusion are important in urban areas, in rural areas they raise particular difficulties due to the distances involved and the lack of transport opportunities available. Whilst an innovative range of coping strategies are in evidence, which often help to build community links (Root, 1996), there are likely to be individuals with very low levels of access to important services and opportunities, and this will have a very real impact on their quality of life. In 1998 nearly 20 per cent of rural households, for example, did not own a car, according to CPRE sources. Whilst there is (rightly) concern for the impact of tax increases on rural car owners, this fifth of the population often appears lost in debates over transport policy direction. Crucially, transport needs to be considered alongside other issues. People themselves rarely separate their lives into the functional segments characterised in government, 'The realities of rural life mean that policies cannot be dealt with in traditional, sectoral ways which, for example, consider education, housing, and transport in isolation from each other' (DOE, 1995c, p. 28).

Reintegrating accessibility issues into governance?

This chapter has highlighted a recent neglect of social issues in transport planning and the marginalisation of transport issues in the considerations of business and state decision-makers. If transport is to be more sustainable, both environmentally and socially, it needs to be integrated, not just into the heart of land-use planning, as is frequently demanded, but into much wider contexts.

The key to this is to think holistically, to see transport planning not just as a set of engineering schemes, but as part of wider processes of meeting people's needs whilst limiting environmental impacts and not adversely affecting local economic competitiveness. Clearly this will not be an easy process, as 'society as a whole is trapped by the automobile. Since lifestyle is the root of the problem, solutions have to be very creative indeed in unpicking the bundle of factors that feed the process of motorisation' (M. Hesse 'Urban space and logistics: on the road to sustainability', *World Transport Policy and Practice* 1, 1995, quoted in Whitelegg, 1997, p. 206).

Change is particularly difficult in the UK in the aftermath of the neo-liberal agenda of the Conservative administrations of the 1980s and 1990s. The marketisation of public services, the fragmentation of governance, the hollowing out of local and national states (Hogwood, 1997) all make the task of any state-led attempt to promote such lifestyle shifts difficult. In relation to transport, two elements of the neo-liberal agenda are particularly pertinent. First, the promotion of 'personal freedom' was a central feature of the 'new right' approach. In transport terms this translated as an expectation of unhindered mobility by car, the freedom to drive a car without consideration of the wider societal implications. Second, Conservative administrations saw road-building as inextricably linked to economic competitiveness. This view, one of 'roads for prosperity' (as the 1989 road-building programme was called) is deeply embedded in the minds of local politicians and others. These two discourses heavily frame the practice of transport planning and the expectations of unhindered mobility prevalent amongst the general public. The 'needs' discourse inherent in welfarist governance viewpoints has largely been displaced. Change will require high levels of understanding and commitment from many in governance, at all levels, in several policy sectors, and such commitment promoted amongst a wide network of other stakeholders.

This is, of course, very easy to say but what are the possible ways forward in addressing current inequities in transport provision? The problems of global equity in the face of international economic and political forces require attention at different levels. At more local levels in the relation of accessibility to opportunities the solutions are evident in the four elements presented by Banister (1997), quoted earlier in this chapter: reducing the distances between people and facilities; reallocating existing roadspace to prioritise public transport, cycling and walking; better integration of public transport modes to achieve a seamless journey; and appeals to end users (principally car drivers) to take a greater social and environmental responsibility. We might add to these subsidy for a greater provision of public transport, particularly in rural areas. Such funding might be derived from an extension of the idea that transport users pay the full cost of their journeys where they are able to do so, the removal of tax advantages for company car users and a requirement for employers to address accessibility issues in the journey to work, perhaps alongside green commuter plans and the remote accessing of services via telecommunications. In addition, there are a plethora of initiatives that can be taken forward by communities, such as social car-sharing schemes, although the very communities that need such schemes are often those without the skills to facilitate them: 'vulnerable groups may become more reliant on the state and the voluntary sector' for their movement needs (Hine, forthcoming). Therefore there may well be a facilitation role for local authorities here.

Central government retains a great deal of regulatory power in transport matters. Its control over finance gives it strong leverage over local authorities to make them toe a particular line. Central government needs to support both policy strategies that demonstrate the equity aspects inherent in all transport policy decisions and initiatives such as concessionary fare schemes that attempt to maintain accessibility and quality of life for vulnerable sections of society. Thus whilst there is a need for change in central government's approaches to these issues, considerable emphasis also needs to be placed on local levels, and on the quality of life of people in their localities. As much as anything else, this suggests a commitment to the integration of policy across different sectors, so that the efforts of area focused initiatives do not conflict with other initiatives: while the concern is for everyone with accessibility problems, the principal focus must be on areas of social exclusion.

Such an agenda suggests the need for funds, flexibility and commitment at the local level, and a central government framework that promotes such an agenda whilst providing the flexibility for localities to achieve it. Such a shift does not, then, imply massive government intervention. Nor does it necessarily require the wholesale integration of government and service delivery, but it does suggest a promotion of the implications of the place of essential land uses, amenities and facilities in the minds of policy-makers in transport and other policy sectors. Initiatives such as the creation of a Women's Unit and a social exclusion unit to cut across traditional government policy sectors is one such way this can be achieved, and the input of these is beginning to have an effect on transport planning within the DETR. In addition, a cross-departmental committee within government is already charged with looking at environmental issues across government departments. The social remit of this committee could be extended to promote the strong equity component that exists alongside the environmental issues in the Local Agenda 21 process.

Whilst local government will play a key role in promoting a social agenda, it is not necessarily the key actor in promoting accessibility issues, and in any case its powers in relation to financing initiatives and leading change are strictly limited. The deregulation of transport-service provision and the spread of responsibilities for transport planning and other services means that local government will have to find a way through the 'multiform maze' of governance (Rhodes, 1997) to achieve change in localities. Local government has lost virtually all direct control over public transport and is hamstrung in attempts at partnership with providers because of the latter's focus on key profitable areas and local government's own lack of finance.

Despite this, perhaps the place to start reintegrating accessibility concerns into the heart of governance would be with local transport policy communities. The policy terrain has shifted a long way in the 1990s, prompted by environmental considerations and public spending cuts. This policy shift is towards a more equitable set of policies too. However, the social agenda still needs to be actively promoted; it hasn't been the driving force for change in contemporary transport policy and the role of transport in processes of social exclusion is all too real. Notions of accessibility have been present in local government for some time but it still appears to be a narrow discourse focusing on making public spaces, buildings and public transport more accessible for the mobility impaired. As a result, the promotion of accessibility issues frequently emanates from a different local authority department from that concerned with transport planning. In addition, responsibility for such issues is usually promoted by a single officer devoted to them. It thus tends to be an add-on to policy, introduced at a particular stage of the design process, rather than the impetus for transport policy as a whole. A social agenda for transport planning would put accessibility issues centre stage.

Conclusions: the road ahead?

There are encouraging signs in the UK that after many decades of transport planning effectively meaning 'roads planning' with other modes marginalised in the activities of transport planning departments, the need for change has been realised. This is the result partly of transport planners having to do things differently following Treasury spending cuts, partly of the environmental agenda gaining leverage over public policy. The new approach to transport planning should be a more equitable transport planning also as, 'solving the pressing problems of transport pollution can also help to solve the recognised inequalities in access to transport' (Barton, 1998a, p. 138). This is not the full picture, however. Whilst many local authorities are putting accessibility issues centre stage in transport planning, there is still a real danger that these issues will remain marginalised in local transport policy and decision-making, with real consequences for the quality of life of the 'accessibility poor'.

Transport policy communities are slowly coming round to the notion of change but a more environmentally and socially equitable transport policy is likely to be highly problematic in implementation terms. Such a transport policy will require effecting changes both in the lifestyles of the travelling public and in the attitudes and mindsets of state and business decision-makers. Both of these changes are likely to be difficult. It is important that an awareness of the consequences of mobility impairment and the needs of the 'accessibility poor' is promoted alongside the environmental issues currently occupying centre stage in transport policy change. Such accessibility issues cut across business, and particularly state, decision-making. In short, accessibility is not just an issue for transport planners.

8 Gender, race and culture in the urban built environment

Ann de Graft-Johnson

Introduction

This chapter comprises a personal account of the author's experiences of social inclusion and/or exclusion as a Black woman architect, working in the world of the built environment professions. It is not intended to be a comprehensive historical account, but seeks to highlight issues of concern. First, the international context is discussed, followed by the European situation; in each case the problem is illustrated by accounts of attitudes and selected instances of exclusion, and then pointers for change are given. Then the chapter focuses upon Britain, highlighting a selection of issues which led to the questioning of post-war reconstruction planning, from both a gender and race perspective. Particular reference is given to the work, publications and research undertaken by Matrix, a highly influential but now disbanded, all women's architectural co-operative practice (see the main Matrix publications, 1979–97, listed in the bibliography). In the last section the question of accommodating diversity in respect of current approaches to professional practice is illustrated. Built environment education is reflected upon in the concluding section, as it offers a means of creating cultural change in the future.

Terminology

Definitions of ethnicity can be problematical. The Commission for Racial Equality (CRE) has stated that ethnic origin is about 'colour and broad ethnic group' and is 'not about nationality, place of birth or citizenship' (CRE, 1991). In Britain the term 'Black' has been used in a number of ways: for instance, to describe people of African descent and, alternatively, as a political description of all groups who face discrimination on the grounds of colour (and, the editor notes, even of Turkish and other Eastern and Mediterranean European peoples). The latter has proved problematic in Britain and there has been a return to more separate definitions for different groups, following more closely the American model. This chapter uses a number of terms and phrases. Where possible, specific ethnicity is given, elsewhere, the following general terms are used:

- White: Caucasian;
- Black: where a general group of African descent is being described;
- Black-African;
- Black-Caribbean;
- Asian;
- Minority Ethnic groups: refers generally to groups, including White people, who are discriminated against because of their ethnicity, such as Irish, Turkish and Kurdish people;
- People of colour: a general term to apply to all people discriminated against on grounds of colour.

'Black' and 'White' are given initial capitals intentionally, in the same way that different nationalities (such as the French) are normally given capitals.

Global hierarchies and the politics of omission

The international context

All the Women are White, All the Blacks are Men, but Some of Us are Brave (Hull *et al.*, 1982). The title of this book summarises the prevalence of an attitude which gives little weight to the specific experiences and contribution of Black women within feminist and Black studies. Even the record of women by women has been marked by omission and the world of architecture and planning is no different. For instance, Claire Lorenz in *Women and Architecture* (Lorenz, 1990) professed to look at contemporary women architects and their practices on a worldwide basis: however, the book is overwhelmingly dominated by White women. Women of African descent are totally ignored, and consequently prominent women architects who more than meet the book's stated criteria, such as Norma Sklarek, are excluded. She was the first woman of African–American descent to register in the United States (New York in 1954 and California in 1962). She was also the first to receive a fellowship from the American Institute of Architects in 1966. She has held high ranking positions in a number of large firms, including that of director and principal and was a founding partner of the all-women partnership Siegal Sklarek-Diamond in 1985. The only women profiled in the book from Africa are White. A very rapid telephone investigation in 1991 revealed women architects such as Mrs Ejiwunmi of Nigeria and Adila Bashir of Kenya in senior positions in their respective ministries of works.

> How many White women have taken on the responsibility to educate themselves about Third World people, their history, their culture? How many White women really think about the stereotypes they retain as truth about women of colour?
>
> (Woo, 1981, p. 143)

At an international conference in Kenya on Gender and Urbanisation in 1994, divisions began to emerge early on between women from developed countries and those from developing countries (Mazingira Institute, 1994). A starting point, to identifying these divisions, was the tabling of potential differences in attitudes to feminism by women from the developing countries. These women, for instance, stressed the need within their cultures to work towards gender equality with (rather than against) men, and they perceived the Western feminist stance to be antipathetic. What was disturbing at this conference was that a number of the women (White) from developed countries were highly dismissive of the other women (Black) and the lessons they could learn from them. This was despite the fact that it was obvious that the women from developing countries often had more to offer in advancing gender-focused strategies than the Western women, because they combined both academic expertise and knowledge with directly applied work at local, grassroots level. These women had a sound knowledge of Western cultures, as well as of their own, and of the history and implications of Empire and colonialism. This gave them a much broader, global overview. For instance, one of the papers presented included research combined with drawn schemes focusing on all aspects of women's lives and cultural contexts. The proposal provided for an integrated, modern, combined housing and workplace development that could be used to advance more general architectural and planning theories and discussion about gender issues in building contexts. Happily, the refusal of women at this conference to go unheard meant that in the end conference feedback and out-comes incorporated positive strategy recommendations that took account of diversity. Proposals included a 'bottom-up', grassroots approach to development and gender aware training and education.

Global governance and change

As the end of the millennium approaches, global aspects of the development of the environment are acknowledged in the political arena as being of major impor-tance. Decisions at global level with regard to development, embodied within United Nations Agenda 21 (Keating, 1993), are having an increasing impact on the local built environment. Although there are large variations between the com-mitments of individual nations, the broad terms, internationally agreed, are the need for better environmental controls, sustainability and equality. The scope of issues involved means that it is essential that integrated, international strategies work at local level. Ever-increasing complexity in terms of the number of relevant issues makes the exchange between the parties involved in producing the built environment, from politicians, professionals and practitioners to end users, cru-cially important. While the need for environmental controls, such as sustainability and lower fuel emissions, makes general sense, it is in terms of impact at the local, applied level that detailed factors, such as who the decision-makers are, become critically significant. The global framework, while on one hand offering the potential to tap the experience of a vast range of practitioners with diverse experience and knowledge, on the other hand produces the problem that decision-making tends to follow the pattern made by the patriarchal and racial hierarchies of world powers. An example of this hierarchy is instanced by one of the United Nations' key organisations, the Security Council, which at present has only fifteen members, ten of them non-permanent with a two-year maximum term of office. The permanent members are the USA, the Commonwealth of Independent States (the former Soviet Union), China, France, and Great Britain and Northern Ireland. There are therefore no permanent members from the African continent or the Indian subcontinent, nor from South America, the Arab states or the Antipodes. The power of veto creates a further imbalance. The imbalance on the world stage moves down the ladder to the cultural bias of those who make the decisions at national level and then down from there to local level. For instance, while the highest levels of consumption and emissions are firmly linked to the dominant powers, in particular the USA, it is the developing countries that are being expected to 'subsidise' others by taking a greater role in achieving reductions.

The breakup of the Soviet Union has put the USA in a unique position as the primary world power. At a European level this has had a huge impact, not only in terms of Eastern European stability but also in terms of human migration and demographics Europe-wide. Added to this, the expansion of the European Community, coupled with the development of the 'fortress Europe' stance has affected non-European immigration and employment and attitudes around race. The tightening of immigration controls in a number of European countries, the growth of the National Front in France and the increased Europe wide network of fascists are aspects of this.

While there are more women participating on the world stage in decision-making capacities, the profile is still overwhelmingly male, with White males predominating, certainly in power terms if not numerically. For example, the USA has an unprecedented say in enforcing sanctions against other countries such as Cuba. Disparities on the global stage are being echoed with the marginalisation of women of colour in mainstream feminism. In general, the global debate and implementation level has tended to be marked by the lack of gender and race equality perspectives and, in the feminist arenas, by an exclusivity which includes race. The lack of acknowledgement of gender issues within policy-making means that decisions taken at summit level may not take account of the realities of daily life or reflect the needs of women. This is despite the fact that women constitute over half the world's population. The raced hierarchy within mainstream and feminist spheres doubly counts against women of colour. All of these factors have implications for policy-making around the urban environment.

The European picture

Many White feminists' failure to acknowledge the differences between themselves and Black and Third World women has contributed to the predominantly Eurocentric and ethnocentric theories of women's oppression.

(Amos and Parmar, 1997, p. 55)

Women of color have been forced to continue arguing for the same basic issues in feminist political organizing, women's studies courses and conferences. Issues of representation, accountability, responsibility, and equal sharing of power and control continue to be major problems in feminist organizing.

(Russo *et al.*, 1991, p. 301)

Within the European framework, as within the global one, there is a pronounced reluctance to discuss and take on board issues of race and culture when developing policy for the built environment, or to look at projects that incorporate diversity. It appears, however, to be even more difficult to overcome this reluctance within the European context, not least because of the lack of representation at conferences. For instance, the author has been the only woman of colour at a number of conferences and seminars. Whether the problem arises from hostility, apathy, lack of knowledge, or resistance to addressing those issues is not clear.

In Britain, at least, we have seen more recently a higher profile for Black and Minority Ethnic women. For instance, Margaret Courtney-Clarke's books *Ndebele: The Art of an African Tribe* (1986) and *African Canvas* (1990) reveal the work of South and West African women. Much of the coverage has focused on traditional lifestyles and work, however, and other work in formulating theories and practice in building modern environments has largely been overlooked.

The policies of different European countries on race and immigration are in part reflected in terms of the degree to which issues of race and culture are acknowledged and discussed within the women's movements of those countries. For instance, German women architects from the organisation Feminist Organisation of Planners and Architects (FOPA) and architectural students invited Matrix, a London-based women's architectural co-operative, to West Berlin shortly before the demise of the Berlin Wall in 1989. (More will be said about Matrix later in this chapter.) A number of projects were visited. These included a women's housing project and a social housing scheme for Turkish people. At the women's housing project one of the host architects, who was also a resident, was asked whether there were any children. She replied that there were none. Shortly after this she said that the only Turkish people in the scheme (a family whose tenancy had been secured by virtue of prior residency) were very lucky to have this accommodation, her tone indicating that she did not believe they deserved to be there. During the conversation two children had been playing in the garden, and it became apparent that because these children were from the Turkish family they had been rendered 'invisible' by this woman. The visit to the Turkish housing scheme offered perhaps the starkest illustration of the degree of separation that exists between the experiences and attitudes of the German and Turkish communities. On a wall was written 'tourists are terrorists', a reaction to the many voyeuristic visitors (including Matrix) to these social housing schemes. Although not universal, the extent of exclusion and the extent to which the feminist debate focuses on White middle class women to the exclusion of others is of continuing concern.

Exclusive attitudes do not just affect discourse between White women and women of colour. A north–south European divide has also been highlighted on a number of occasions. At a conference in the Netherlands in 1994, '*Emancipation as Related to Physical Planning, Housing and Mobility*', the lack of clarification of terminology meant that contradictions often arose (Matrix, 1994a; Ottes *et al.*, 1995). Terms such as 'mobility' may apply to social, economic, transport mobility or mobility in relation to disabilities; similarly, 'access' has broad and differing interpretations both in local and global debates, being linked to disabilities, employment or education. The author was advised that there were exemplars that could have illustrated more culturally diverse approaches to gender issues, but that these had been largely excluded. Policy recommendations tended to favour northern European conditions to the exclusion of factors affecting some of the southern European countries, both in terms of definitions and in terms of detail. This led to some women being considerably disenchanted with the conference outcomes. For instance, recommendations on transport focused on the bicycle, ignoring issues of topography, demography, climate and culture. The simplistic, prescriptive, monochromatic ethos of this conference meant that an opportunity was lost to develop more enlightened discussions and strategies that would have allowed for adaptation to particular localities, to ensure their appropriateness. A greater focus on wider-ranging cultural issues could and would have resulted in better policy frameworks and recommendations.

Britain: sustainability, participation and equality

The same but different

This section looks more closely at the situation at a local level in Britain, in order to highlight connections with the global and European picture and to show more specifically the effects and implications of policy-making. There are both similarities and dissimilarities between Britain and other European countries, not least within historical, economic and social frameworks and suffrage. For instance Black and Minority Ethnic people in Britain can become citizens, whereas in countries such as Germany second- and even third-generation Minority Ethnic residents cannot achieve this (being treated as guestworkers). However, the use of specific national issues may flag up principles or concerns that may be more broadly applied.

In this section the concept of the 'successful city' as perceived by post-war planners will be set against the issues raised by minority groups. It is difficult to define what one considers to be successful cities without becoming overly nostalgic or sentimental, but all successful cities have a degree of diversity and potential for development which reflect participation by citizens. Obviously, large cities such as London are different in their composition from smaller cities with less complex overall structures, but most, nowadays, contain an element of diversity and change.

Historical context

The Industrial Revolution in Britain marked the start of enormous changes to the demographic profile of the country and caused the rapid growth of urban living (Clare, 1994). While social housing provision in the form of almshouses and workers' cottages goes back several centuries, the nineteenth and more particularly the twentieth century saw a huge increase in housing provision for working-class people and a greater input of planning and architectural theories in developing housing schemes. Much of this accumulated urban development remained unchanged until the mid-twentieth century, when a new comprehensive national town planning system was introduced in 1947 after World War II.

Post-war reconstruction decanting policies in the 1950s and 1960s, together with housing development on a massive scale fed into the dismantling of the extended family networks that were previously commonplace, certainly among the working classes. Many residents professed themselves happy with the new idealistic schemes when they were first built (Roberts, 1991; Gilroy and Woods, 1994). However, various factors, including issues of management (both social and maintenance), lack of linked amenities such as shops, social and recreational facilities and also of potential for informal social interaction within the design of the schemes, saw a relatively rapid decline of many of the projects. Not only did high rise schemes become failures, but also, in some cases, so did more conventional housing projects. Local newspaper articles both pre- and post-war, for instance, in Tower Hamlets, London, indicated levels of concern about the impact and wisdom of such large-scale redevelopment which inflicted more demolition than the Blitz (Gilroy and Woods, 1994).

Reaction

> Cities are fantastically dynamic places, and this is strikingly true of their successful parts, which offer a fertile ground for the plans of thousands of people.
>
> (Jacobs, 1964, p. 24)

Approaches to city development which assume a homogeneous population (as in the post-war period) and a zoned series of functions lack the integration of elements that might achieve more 'successful cities'. Cities which have no human investment or participation from their populations, but are solely products of external design and input are unlikely to create the framework for a flourishing environment.

Changes in architectural and planning theories have had a major impact on the built environment in this century. By the 1960s reaction had set in. Jane Jacobs stated in *The Death and Life of Great American Cities* (Jacobs, 1964) that modern city planning and architecture were directly linked to Ebenezer Howard's garden city principles (Howard, 1898). Intrinsic to his theories was the segregation of use, so, for instance, industry and commerce were separated from housing, schools and green areas. Jacobs described these principles as 'anti-city'. She believed that they militated towards a zoned, static, inflexible, municipally controlled environment which was essentially destructive of the city fabric and the elements ensuring its success and diversity. Howard's ideas were adopted by architects and planners who developed them as decentrist strategies. Jacobs argued that the notions of zoning and attitudes to the garden or 'park' were adopted by Le Corbusier and that modernism grew directly out of garden city principles – but with quite different results.

> Le Corbusier accepted the Garden City's fundamental image, superficially at least, and worked to make it practical for high densities. He described his creation as the Garden City made attainable.
>
> (Jacobs, 1964, p. 32)

Factors of development must not only incorporate the notions of sustainability and equality but also processes that are representative and at a workable scale to realistically allow input by all sectors of the society. By the 1970s, reactions to the post-war reconstruction period of 'town planning' were becoming less generalised, more gendered, and specific to the needs of different racial and cultural groups.

> Routes, design of spaces, their allocation and location often reflect the interests of the dominant sex, cultural or racial group without regard to the needs and expectations of those who are not in decision-making positions.
>
> (de Graft-Johnson, 1991, p. 9)

Women's domain and the emergence of Matrix

Changes in theories affecting urban design, transport and the domestic and working environment have historically taken place without input from women. Features in urban design and transport policy have often acted against women, both in their own right and as primary carers. Access, including access for the disabled and pushchairs, has often been overlooked, not considered important or even deemed detrimental to design (Matrix, 1984). In Britain transport policy and related design have favoured the car rather than pedestrians or public transport. Men have historically monopolised car use to the exclusion of women; the majority of users of public transport have been women. Provision for the latter means of transport was and is perceived by many women as neither reliable, safe, comfortable nor suitable, particularly when they are with small children. While surveys indicate that it is young adult males who are most likely to be attacked in public spaces, the self-imposed curfew that women often place on themselves may mean that women's lesser use of the public domain affects the statistics so that they do not reflect actual potential risk. Because of their increasing choices, women have opted for cars in large numbers and the more recent substantial increases in car usage have been identified as being almost entirely attributable to women adopting this form of transport. The number of women in Britain passing their driving tests has more than doubled over the last twenty years (cf. ONS, 1996, Table 12.8; 'Cars and car ownership by region'; and chapter 7 above). Car manufacturers are increasingly targeting their advertising at women and have confirmed that the recent upsurge in car purchases in the late 1990s is almost entirely attributable to women purchasers. The benefits are obvious. Cars can be driven right from one's door to the destination so that there is less exposure to potential dangers. They facilitate multi-purpose journeys and complex scheduling, including dropping children off to school, going to work, the supermarket and to leisure activities. The alternatives: walking, bussing, the underground or train are often perceived to be – or are – unsafe, inefficient in terms of time and a source of access problems. Similar factors or worse face Black and Minority Ethnic people.

It is not only in the public domain that gender issues are a factor. In the home often women have had no space of their own and no rights to privacy. Moreover, they are just as likely, or more likely, to face violence there as in public spaces. The development of women's rights and refuge organisations has raised the profile of abuse and violence against women and children within both the public and the domestic sphere. Women's engagement in the political arena has also included critiquing aspects of the built environment and challenging attitudes that have worked against or ignored factors of women's lives.

One of the groups in Britain that emerged out of concerns about built space, attitudes to women in the building professions and policy was Matrix. In 1978 a feminist architects' group was formed, entitled the Feminist Design Collective. It was originally a discussion group which also undertook architectural work. In March 1979 a conference called 'Women and Space' was organised by this group. In 1980 the Collective split and an umbrella organisation, Matrix, was formed to work on various projects. These included work for women's groups, an exhibition called 'Home Truths' (Matrix, 1980), conference presentations (Matrix, 1979, 1994b, *inter alia*), the book *Making Space* (Matrix, 1984) and a range of commissioned research (Matrix, 1993, 1996 *inter alia*), all of which was published under the group's name; contributors to the collective's publications, research and design projects were not named individually. The book covered issues such as public space, housing and women's experience of the built environment and the architectural profession. It also broached different ways of working, and examined feminist approaches to environmental design which took account of safety, access and the realities of women's daily lives. In 1981 some of the members went on to become founders of Matrix Feminist Architectural Co-operative. They aimed to put theories into practice and develop new processes to ensure inclusivity, not just in relation to gender issues. As a practice Matrix sought to develop participatory methodology that would ensure full, informed participation of clients and users and to prioritise projects that benefited women and children.

Matrix has confronted a changing array of challenges. The 1970s also saw the advent of the New Architecture Movement (Matrix, 1984) which advocated a community-led, participatory approach. The development of the housing association movement also contributed towards a change in attitudes and large-scale redevelopment gave way to smaller scale refurbishment and regeneration. However, recent years have seen a return to large scale urban regeneration and development projects, often with consortia development teams. The current development of postmodernism in this country has done little to reappraise attitudes to space and has generally been one of 'facadism' and 'pastiche' which may be viewed as being culturally stultifying.

Feminism and race

Twentieth-century Britain has seen many changes in women's lives and roles, not least women's suffrage. External factors affecting women's roles have followed tides of participation and non-participation for many women in the workplace. The war years and the period immediately after saw large-scale employment of women in non-traditional gender roles. This went along with different attitudes to family policy, such as provision of children's nurseries. The post-war era marked the return of men to their traditional workplaces, of women 'back to the kitchen' and the domestic domain and the closure of many of the nurseries and provisions which facilitated women's employment (Holdsworth 1988).

A number of factors, including the loss of many traditional areas of male employment such as mining and the growth, on the other hand, of service industries which have employed women more, have led to the large increase in the number of women in paid employment in recent years (ONS, 1996, Table 4.3: 'Economic status of people of working age, by gender and ethnic group'; and annual).

The advance of feminism in the last twenty-five years in this country and longer in the USA has seen challenges to aspects of women's lives and roles. However, many aspects of mainstream feminist argument failed, for instance, to take account of the position and realities of Black and Minority Ethnic women. Black women, for instance, had never left the workplace, often filling low-paid service-sector roles as cleaners or nurses, working antisocial hours. Black women's reproduction issues were often different. Where White women were arguing for the right to have abortions, Black women in the UK, as well as the USA, were pointing out that for them getting abortions was often not the problem: the issue for many Black women was forcible sterilisation. Having rights over one's own body should have been the central argument. Despite the influence of civil rights, independence movements, Black women's participation in and contribution to suffrage and promoting women's rights, the focus of the women's movement has often failed to reflect the differences in women's experience.

> Black women's continued challenges to the question of forced sterilization and the use of the contraceptive drug Depo Provera has meant that such campaigns as the National Abortion Campaign have been forced to reassess the relevance of their single issue focus for the majority of working class Black women and to change the orientation of their campaigns and actions.
>
> (Amos and Parmar, 1997, p. 56, para. 4)

Current practice and realities

Regeneration in practice

In this section current professional practice and planning policy will be reflected upon with particular reference to urban regeneration, and in the following section the situation in built environment education and related professional membership will be discusssed as contributory factors in shaping the planning culture. Housing and regeneration form a major focus for current and future development. Schemes under way or projected are often large scale, consortia structured, involving large sections of the local populace. The briefs for these projects place emphasis on the need for both sustainability and equality within participatory frameworks. Selection processes in areas which, for instance, have high levels of unemployment, low incomes or substantial Black and Minority Ethnic populations highlight the importance of an awareness and understanding of social factors. Development teams routinely state their commitment to, and embracement of, equal opportunities in relation to class, gender, race, disabilities and so on, producing detailed statements of equal opportunities and outlining their methodology for encouraging participation to substantiate this in their bids for commissions. The question is whether or not the actual practice and resultant schemes embody these commitments; also, whether the framework of funding in fact permits the appropriate representative development. This is despite talk of the need to produce integrated and sustainable environments. Sustainability is not just viewed in terms of ecological factors. Equality, participation and investment in the formulation of the environment by all sectors of the population are also viewed as critical factors in achieving sustainability. Scrutiny of practice, the profile of active participants and the specifics of built projects would indicate that often the paper principles of equality and diversity are not achieved in reality. Major schemes have gone ahead in Hulme (Manchester) and Southwark, Lambeth, Lewisham, Hackney, Tower Hamlets and Brent (London), which all have large minority populations. For instance, 57 per cent of the population living on the five Estates Partnership Project in Southwark, London, are of Black or Minority Ethnic origin.

> None of these projects have any significant participation by Black and Minority Ethnic construction professionals. There are no significant monitoring procedures to ensure adherence to equal opportunity policies.
>
> (B. Grant *et al.*, 1996, p. 18, para. 4.2.4)

Although women are increasingly involved in these projects at senior levels, the profile is still predominantly male at the top. Very few Black, Asian or Minority Ethnic people have any input at a directive level. The Joseph Rowntree Foundation has raised concerns that mistakes of the post-war era are being repeated in the new schemes and that they are consequently not achieving the required long term sustainable development (Harrison and Davies, 1995).

The consortia-nature of these projects, together with their scale, creates problems in terms of participation. For instance, lesbians and gays who may have active input within the context of a specialist co-operative or a specific housing project are less likely to feel at ease raising issues within a large project where their visibility might have serious implications in the future.

Housing associations representing Black and Minority Ethnic groups, women, people with disabilities, lesbians and gays are involved in these projects and are more representative of the groups they serve, but they are unlikely to be project leaders. A number of development workers for these housing associations have voiced concerns that in the context of large-scale or consortium projects they are often marginalised by larger leading housing associations. Consequently they have problems in ensuring that the needs of their tenants are taken on. Often residents feel better able to talk directly to someone whom they feel more closely represents their experience and may recognise their needs more directly. Black and Minority Ethnic consultants or women have actively been requested by residents' groups. Yet all too often leadership is White, male and middle class, with any minority team members in marginal positions. In some cases, because of discrimination and White dominance of residents' organisations, Black people may refuse to participate in meetings. There are examples of White male professionals failing to understand or take seriously gender, cultural or racial requirements, for instance security precautions. One example of a design team's lack of awareness is instanced by an Asian housing scheme on which the architect failed to take account of the future residents' average height. The result was that the security spy holes were considerably above the eye levels of the user group. Le Corbusier's modular man was based on the height of a male 1830mm tall with a reach above floor level of 2260mm. The average height of women generally and some ethnic groups in particular is considerably below this.

Various groups and organisations, including the Society of Black Architects (SOBA) have been working towards achieving greater equality for Black, Asian and Minority Ethnic people in the sphere of the built environment. In 1993 a booklet was released, *Accommodating Diversity*, commissioned by the National Federation of Housing Associations (Penoyre & Prasad *et al.*, 1993). Its remit was to identify design issues specific to Minority Ethnic groups. The groups identified constituted groups based in this country, and the booklet sought to highlight considerations such as lifestyle, family structure, religious practice, beliefs and safety without being prescriptive and generating a further set of false stereotypes.

The four architectural practices involved in producing this book were also part of a working group of Black construction professionals chaired by Bernie Grant MP. The work of this group and subsequent report *'Building E=quality'* (B. Grant *et al.*, 1996) highlighted many of the discrepancies between stated practice and reality. The formation of the working party was triggered by the fact that many of the larger scale urban projects were in areas with large numbers of Minority Ethnic residents and there was concern that the needs and wishes of these groups were being overlooked. This was particularly so because organisations with specialist input to represent them were being marginalised by the

Figure 8.1 Pose based on Le Corbusier's Modular Man.

Figure 8.2 Dwelling designed to be used either as a single extended family house or divided into two self-contained homes: a two-bedroom ground-floor flat and a three-bedroom maisonette on the first and second floors. Both homes will have access to a rear garden. Tiller Road, East London, Labo Housing Association (Homeless Minority Ethnic Families: Bengali, Somali, African–Caribbean, Chinese, Vietnamese). Architect: Matrix Feminist Architectural Co-operative. Source: Penoyre & Prasad *et al.* (1993), pp. 14–15. Reproduced by permission of the Home Housing Trust.

Ground floor

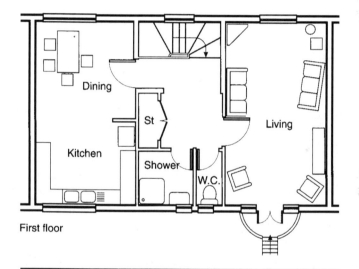

First floor

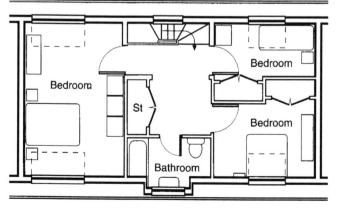

Second floor

Figure 8.3 Bedroom with alcove for child's cot or study space. Vaughan Way, East London, Labo Housing Association (Homeless Minority Ethnic Families: Bengali, Somali, African–Caribbean, Chinese, Vietnamese). Architect: Matrix Feminist Architectural Co-operative.
Source: Penoyre & Prasad *et al.* (1993), p. 37. Reproduced by permission of the Home Housing Trust.

large-scale consortia-nature of the projects and the fact that the profile of the leading professionals involved in no way matched that of the users at the key decision-making levels. A number of problems were identified in this report relating both to Black and Minority Ethnic people as professionals and as users. The funding, and therefore the decision-making powers, were being channelled into projects headed by White, generally male officers. Again, in terms of commissioning the construction teams, Black and Minority Ethnic professionals as well as women had failed to secure roles as leading consultants. The system of approved panel criteria and any other selection practices linked to ensuring equal opportunities was shown to have failed signally.

The house types shown in Figures 8.2 and 8.3 illustrate how accommodation can be designed to meet cultural needs while at the same time offering flexibility.

In 1990 the London Equal Opportunities Federation (LEOF) was founded by fourteen of the larger London housing associations, purportedly to tackle discrimination against women and people with disabilities, as well as with regard to race in the award of contracts. However, a survey of commissions awarded by the associations in 1995 revealed that nine of the fourteen associations had failed to make any awards to Black and Minority Ethnic consultants. Black and Minority Ethnic consultants only won 0.14 per cent of the total fee allocation. In other words, the commissions were excessively below the representational position of the consultants on the list. For example, one of the housing associations had indicated a target of 18 per cent Black and Minority Ethnic representation on the consultants' list, in line with their tenant profile. A number of practices believe that they were on the approved lists to give an indication of commitment to equal opportunities while discriminatory practices were being perpetuated. Projects were still being largely given to White, male led practices whose stated policies on equality might well not have stood up to scrutiny (cf. Matrix, 1993).

Matrix's own experience of having been selected for lists or carried out at-risk work such as feasibility studies, costings and finding sites which failed to achieve promised commissions, indicated on several occasions the levels of prevailing discrimination. Also, at the competitive job-bidding level client expectations of the profile of the professionals fed into this atmosphere of discrimination. On one occasion, two Black women from Matrix went to an interview as potential lead consultants with a White male potential sub-consultant. The chair of the interview panel was informed that Matrix had arrived and was in the waiting area. He emerged from a room and went straight past Matrix, into the corridor – only to reappear as he realised that he had passed Matrix and that Matrix, on this occasion, was Black! Matters did not improve once in the interview. Despite the supposed representative composition of the panel, questions were directed at the White male.

The relevance of the lack of Black and/or female decision-makers, leading consultants and construction professionals becomes clearer when looked at in the context of some of the urban development projects. As previously stated, large urban developments with substantial Black and Minority Ethnic resident populations have gone ahead with government funding. Projects in London such as Stonebridge and Peckham Five Estates, and Clapton and Hulme in Manchester, all have major projects without significant participation by Black and Minority Ethnic professionals. This is despite concerted efforts by residents in a number of instances to push for more representative consultants. Steve Quilley of the University of Manchester commented in an Open University programme *New Forms of Partnership* (OU, 1998) that the Hulme Estate had effectively become a gentrified, private sector, property-led, 24-hour service city to Manchester's centre. He further stated that the design was misconstrued because it did not focus on local employment and it increased social polarisation. Residents of Tiger Bay have raised similar concerns on local broadcasts in relation to the Tiger Bay development in Cardiff, namely that the projects put more emphasis on the commercial sector than on the needs and wishes of the original population.

> Urban regeneration programmes could give an unprecedented opportunity to try to ensure that jobs are created and the profits made are put to the benefit of local communities, many of which, Black and Minority Ethnic and White alike, suffer exceptionally high levels of social deprivation.
>
> (B. Grant *et al*, 1996, p. 25, para. 4.7.1)

Matrix bade for one large commission with obvious backing by tenant representatives. After many hurdles, including answering satisfactorily permutations of the same questions several times over, demonstrating, for instance, that Matrix had an adequate office, the commission was awarded elsewhere. A fellow architect remarked 'Oh yes, I heard it went to a man and a dog in a basement!'

Obviously the success of schemes is in part dependent on the commitment, awareness and sensitivity of the design team. A starting point for much of this is the standard of the methodology for determining inclusion in the consultation process. This can become particularly critical for groups where English is a second language. Often actions taken can isolate or marginalise. For instance, on one project involving four estates with different issues, initially meetings were held for Kurdish/Turkish people from all the estates. The result was that it was impossible to deal with the specific issues on their respective estates and they felt excluded from the main discussions held with the other people on their own estate. When they then decided to attend the meetings for their estates backed by their interpreter, adequate time and space was not afforded for translation and their input.

An interesting comparison between attitudes to development arose a few years ago from two initially very similar projects in Hackney involving high-rise blocks. The pilot scheme, which went ahead with full residents' participation, resulted in various improvements such as the provision of a 24-hour concierge with resident-focused management and provision for the elderly, together with social amenities on an intermediate floor rather than at ground-floor level. The council noted the success of the project and moved on to another high-rise building. However, it failed to take on board the critical key element of the previous project, namely participation, and instigated 'improvements' to the amenities without consultation. The council representative interviewed described with pride the management/ computer technology aspect of the project and, even when pressed, failed to understand the feeling of investment that proper participation in decision-making by residents achieves.

The way in which funding is allocated on projects has a strong bearing on the development of the urban fabric. Zoning of funding militates towards equivalent zoning of the urban environment. This means that provision of necessary linked amenities such as shops, childcare facilities, training and provision for youth is not included within the main project and is first dependent on being recognised as necessary and then subject to further funding bids, which may or may not be successful (Harrison and Davies, 1995). The lack of funding for integrated development is one of the particular areas identified by the Rowntree Foundation as being obstructive to notions of sustainable development. Additionally partnership money allowing for private and social housing development, rather than producing an integrated approach has sometimes added to divisiveness because the commercial aspect can overtake local economic and social needs, not least with regard to employment and training.

The nature of staffing composition in those organisations responsible for overseeing regeneration is also an important factor. For example, in 1993 Matrix found from its ongoing research that local authority planning departments manifested inequalities in staffing structures in terms of the distribution of women and ethnic minorities, and lack of gender parity. In terms of policy, most councils did not include specific gender issues and many had recently reduced or closed women's units. In general, Unitary Development Plans covered some women's issues non-specifically, in areas such as aspects of harassment safety and security or women as primary carers in relation to the provision of childcare facilities or recreation.

Built environment education

Students

It is obvious to anyone interested in actively furthering the cause of equal opportunities, improving representation and encouraging more progressive attitudes that education and training are key. While schools now do far more to promote career options in terms of offering placements and providing information, serious questions remain about the extent to which subject options still follow assumptions about appropriate careers for girls. Attitudes to Black people in mainstream education have given rise for concern. For Black males this has led to under-achievement, exclusions and higher than average unemployment. For Black females the picture has been different. Dr Heidi Safia Mirza argues in her book *Young, Female and Black* (Mirza, 1992, and see Mirza, 1997) that in the face of discrimination Black females have identified educational routes that they can take and strategic careers. The author's own witness is that this often means that in terms of time-scale, career progression and choice of course, Black women's experience and road to a career are often non-conventional. For instance, a significant number of Black women will not go on to university/further education direct from school. Many will have taken paid work, attended evening courses, possibly had children and then entered degree courses as mature students. The fact that students entering courses covering the built environment are predominantly White and male does indicate that careers in these areas are still not perceived as options by many women and Minority Ethnic groups.

A number of studies have been made of the representation of women and Minority Ethnic groups among students. However, there are questions about the framework for gathering statistics, in terms of the definitions used and their scope. For instance, Matrix research reveals that there are likely to be differences between representation of Black-Caribbean students and those of direct African descent, the numbers of the former being exceptionally low. But few questionnaires bring this to light. In 1996 the Royal Town Planning Institute figures indicated that 43 per cent of their student members were women. Figures produced by the Royal Institute of British Architects (RIBA) in the same year showed lower representation of women on architecture courses, at 31 per cent. Obtaining statistics on ethnicity is problematic because of discrepancies in monitoring procedures in the construction professions and academic institutions. The RIBA noted in its *Education Statistics* 1991–2 that many schools of architecture did not have a policy of recording or providing information on ethnic origin (RIBA, 1992; and cf. WAC, 1993; Matrix, 1993). The RIBA has ceased collecting statistics on ethnicity and the Chartered Institute of Building Services Engineers has never collected data (cf. Greed, 1999). Other statistical information only differentiates between White and non-White and does not provide indicators of the ratio of overseas students or the representation of different ethnic groups. Black and Minority Ethnic groups form approximately 5.8 per cent of the population (ONS, 1997). RIBA figures in 1995 indicated that at entry level the figures were in line with the national average. However, as the 1993 RIBA figures showed only 2 per cent of architects described as other than White British, there is concern that there is a significant dropout rate.

In 1993 Matrix was involved with five other research organisations from Belgium, Greece, Germany, France and Denmark in a participatory research project commissioned by the Commission for European Equality. Part of the remit was to look at the representation of women and attitudes to equality in schools of architecture and planning. All fifty-two schools of architecture and planning were sent questionnaires. Only eleven out of the fifty-two schools replied in writing, although there were some telling telephone conversations. Some schools expressed their unwillingness or inability to reply because relevant statistical information was not held. Matrix research indicated that in architecture more women than men dropped out between part II and the professional qualification level, part III. An indicator of the level of commitment to adequate equal opportunities implementation was one conversation in which a respondent confirmed that although they had an equal opportunities policy, there was no monitoring procedure. When asked how the school was able to assess the success of the policy the response was 'Precisely!' Some members of faculties revealed concerns that there was active support for the ideas and ideology of patriarchy and overt sexism. One staff member stated that there was 'an unfortunate perpetuation of the traditional view of architecture as a male role'. Another comment was that 'a great deal of design work is carried out by young, able-bodied males who do not comprehend that the majority of the users are not similar'. There was a general assumption that positive action and positive discrimination were one and the same thing, and a view that positive action was a form of discrimination that was unacceptable. Positive Action was not viewed as a means of taking steps to ensure non-exclusive access to posts and courses. The author is aware that this confusion persists.

Most of the respondent schools had equal opportunities policies which were directed at both staff and students. But only three included gender studies as part of the course and only five were involved in research in gender issues. Although notions of multiculturalism are appearing in some curricula this is probably in a very limited context. Courses still overwhelmingly focus on White, Western, male architecture and theory. The influence of architecture from other parts of the world fails to be picked up. There are also huge voids in knowledge, in part linked to Empire, colonialism and slavery, which have caused, for example, certain historic African and Asian (Dravidian) architecture and town planning to be largely written out of the curriculum.

In 1990 the Society of Black Architects (SOBA) was formed and has been lobbying for furtherance of equal opportunities in education, training and employment. One of the focuses has been course content, and SOBA has put the case for including world architecture in the curriculum. SOBA has also voiced its concern over the low representation of some groups and the fall-off in representation from schools to employment. One of the concerns is the very low representation of groups such as Black-Caribbean UK-based students, particularly where there is a high concentration of Black people in the local population and the particular institution has local students from other ethnic groups.

Academic staff

Teaching ratios identified in research undertaken in 1993 by Matrix were neither representative of the national population nor of the student constituency (Matrix, 1993, 1994b, 1996). In 1992 women constituted 44 per cent of a total national work force of around 25 million and represented 32 per cent of people in full-time employment (according to the Equal Opportunities Commission, who provided Matrix with information from OPCS, 1992, Tables 1 and 2). These percentages have since increased. By contrast, the ratios of female teaching staff within planning schools and architecture schools were 9.5 per cent and 17 per cent for full-time posts respectively. The ratios increased to 15 per cent and 25 per cent for part-time, temporary contracts. All faculties who replied to the questionnaire had male directors, deans and deputy directors, although there were a number of women professors. Planning and architecture schools were not representative of the national profile, nor were they of other areas of employment in faculties. Women were concentrated in administration and as library staff, representing 85 per cent and 70 per cent of the totals for these respective areas.

The construction professions

As can be seen from Figure 8.4, there are significant variations in the representation of women in the construction disciplines. In 1996 the Chartered Institute of Housing (CIH) had the highest representation of women, at 47 per cent. This discipline may also have a high percentage of Black and Minority Ethnic members (Greed, 1998). The Chartered Institute of Building (CIOB) had the lowest percentage of women at 0.9 per cent. In architecture and planning the figures were 11.5 per cent and 22 per cent respectively (Greed, 1998).

Figures on ethnicity are not available across the professions. In 1995 Matrix produced a report, *Women in Architecture*, funded by the Arts Council (Matrix, 1996). This report (not to be confused with the report of the same name by the RIBA Women Architects' Committee; see WAC, 1993) was based on findings from a questionnaire sent to all registered women architects in the UK. In 1995 the total number of architects was 30,442, of whom 2,829 (9.3 per cent) were recorded by the Architects' Registration Board (previously ARCUK) as female. There were 1,019 replies from women to the Matrix questionnaire, a response rate of 36 per cent. A question on ethnicity was included which 966 women answered. This survey, whilst limited, did give some indication of disparities between representation in the population at large and within the profession (see Tables 8.1 and 8.2). Of the main ethnic groups listed in census

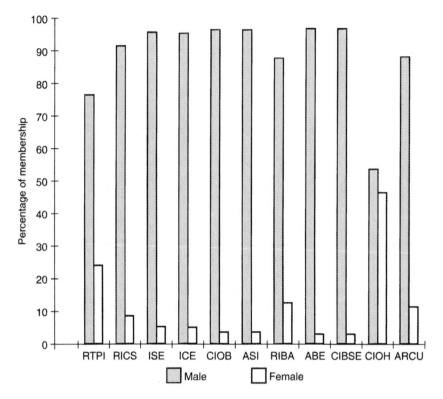

Source: Data from research by Clara H. Greed; see Table 15.1 below.

Figure 8.4 Construction professions: total membership.

Table 8.1 Population by ethnic group

	Millions	*% of population*
White	52.9	94.0
Black		
Black-Caribbean	0.5	0.9
Black-African	0.3	0.5
Other Black groups	0.1	0.2
Indian	0.9	1.6
Pakistani	0.6	1.1
Bangladeshi	0.2	0.4
Chinese	0.1	0.2
None of above	0.7	1.2
Total	56.3	100.0

Source: ONS (1998), Table 1.10: 'Population by ethnic group and age'.

Table 8.2 Ethnic breakdown of women architects, 1995

Respondent self-definition	No. of respondents	% of respondents
White	889	92.03
Black-Caribbean	4	0.41
Black-African	9	0.93
Black other	2	0.21
Indian	5	0.52
Pakistani	1	0.10
Bangladeshi	1	0.10
Chinese	19	1.97
Iranian	2	0.21
Jewish	3	0.31
Irish	7	0.72
Mixed blood	1	0.10
Persian	1	0.10
Anglo-Arabian	2	0.21
Greek/Cypriot	5	0.52
Scottish	2	0.21
Welsh	1	0.10
Euro-Asian	2	0.21
Sri Lankan	1	0.10
Eastern European	1	0.10
White-Caribbean	1	0.10
Czechoslovakian	1	0.10
Malaysian	1	0.10
White-African	2	0.21
Pink, etc.	1	0.10
Russian Anglo-Indian	1	0.10
Australian	1	0.10
Total	966	100.00
No response	53	5.20

Source: Matrix (1996)

Note
Total estimated number of women architects: 2,829
Total number of architects, male and female: 30,442

figures, only women architects of Chinese descent would appear to be well represented in the architectural profession as a whole in relation to their profile in the population. None of the other main groups achieved their representation in the population at large as architects. Black-Caribbean, Indian, Pakistani and Bangladeshi women architects were very poorly represented. Clearly there is need for both numerical and cultural change in the composition of the construction professions, as also discussed by Huw Thomas in chapter 2, and further by Clara Greed in chapter 15.

Local authorities

Matrix research findings in 1993 on local authority planning departments again highlighted inequalities of distribution and lack of gender parity in staffing.

Conclusion

The author has sought to identify a selection of the complex social, economic and educational factors which impact on the planning and design of the built environment. Through each stratum, from local grassroots level to the global political stage, questions still have to be asked about the gender and complexion of the decision-makers of today and potential ones of tomorrow. Despite the stated notions of equality and participation embodied in UN Agenda 21, which has been adopted in many regions, much of the actual deterministic profile follows another path, a hierarchical one that is both gendered and raced. Large scale regeneration projects such as Stonebridge in London and Hulme in Manchester have demonstrated, pre-development, the strength of local residents' wishes to determine their environment, and their understanding of the crucial link between employment and residence to ensure success. This aspect can be reflected worldwide. Critiques have already been produced on the Hulme development, questioning whether the regeneration has addressed the requirements of the original residents. Stonebridge is still being developed. However, it is clear, irrespective of success or not, that the current situation is too ad hoc and that the policy framework needs to incorporate evaluation and monitoring to ensure accountability as an integral part of any development policy which aims at inclusion, participation and equality, at whatever level, whether global or local.

Part III

New policy horizons

9 The European Union

Social cohesion and social town planning

Richard Williams

> Social Policy is central to the process of European integration.
> (Padraig Flynn, member of the European Commission,
> CEC, 1996a, p. 5)

Ebenezer Howard, in his seminal work *Tomorrow: A Peaceful Path to Real Reform* (1898), recognised very clearly the links between town design, regional spatial planning and social policy in his concept for the 'Social City' (Hall, 1992; Ward, 1994). His work generated worldwide interest, and his ideas influenced the development of planning thought in many other European Union (EU) member-states besides the UK. Now, one hundred years later, the question to be explored here is whether this fundamental link between social policy and what is now to be termed 'spatial planning', is on its way back to the UK, in ways that may not be readily recognised by British town planners, through the policies and programmes of the EU.

Scope and structure

The central purpose of this chapter is to discuss how far the social planning agenda in the UK, as defined and discussed earlier, is shaped or framed by the EU context and EU initiatives and programmes. The starting point is the key phrase 'economic and social cohesion'. This is found in the first of the objectives of the Union as set out in Article 2 of the 1997 Treaty of Amsterdam:

> The Union shall set itself the following objectives:
> * to promote economic and social progress and *a high level of employment and to achieve balanced and sustainable development*, in particular through the creation of an area without internal frontiers, through the strengthening of economic and social cohesion and through the establishment of economic and monetary union, ultimately including a single currency in accordance with the provisions of this Treaty [continues]
>
> House of Commons, (1997a, p. 113); emphasis added.

Article 2 expresses the fundamental goals of the EU. The words in italic indicate amendments added in the Amsterdam Treaty to the form of words adopted in the Maastricht Treaty. The phrase 'economic and social cohesion' occurs in many EU contexts, as does its counterpart 'social exclusion'. This chapter aims to offer an answer to the questions of what is the meaning of the word 'social' within this phrase, and how might this impact on town planning and contribute to the practice of social town planning.

Chapter 1 refers to the various spatial scales at which social town planning can be considered. The starting point for this present chapter is the largest spatial scale of jurisdiction affecting local and regional authorities, namely the EU. Although this jurisdiction does not extend explicitly to town planning law, it is intended to show that the EU does have relevance at the more local spatial scales at which social town planning initiatives are normally to be expected, namely those of the city and its neighbourhoods.

The social dimension of the EU generates its share of rhetoric, as the quotation from Commissioner Padraig Flynn illustrates. He is the Commissioner responsible for employment and social affairs, and for relations with the Economic and Social Committee. The underlying paradigm, a product of the social democratic consensus that has been a feature of the formative years of the EU, is the need to balance economic development with social programmes, in the spirit of social partnership. The aim is summed up in the key phrase from Article 2 which is repeated frequently in statements of policy objectives: 'economic and social cohesion'.

The rhetoric of European integration, and the thought-processes that lie behind the generation of tangible policies and programmes of actions from grand words and rhetorical concepts, are often a mystery to the pragmatic British mentality. The consequence is that there is a tendency to takes one's eye off the ball: one may then be taken by surprise when a new policy is presented by the EU Council of Ministers for approval, or one may fail to recognise a programme for what it is and risk missing opportunities that it may offer.

This chapter has, therefore, the twin tasks of, first, setting out the scope and range of the social programmes and social cohesion policies of the EU, and their legal basis and, second, relating this to the activities of local planning authorities and to EU intervention in the field of spatial planning. It thus becomes possible to review the social agenda of the EU and examine how far this influences, or is to any extent implemented by, the spatial planning agenda.

The following structure is adopted. The first section discusses the place of social policy within the EU as a whole, and within the treaties. Within this section, the issue of terminology is specifically addressed, since in this field, as in many others, understanding can be impeded if the assumption is made that EU usage of English words always carries the same senses as UK usage. This section is followed by a systematic review of the main components of EU social policy that are of interest in the context of social town planning, concentrating on the structural funds, the single market, cohesion policy and the social chapter. Both their overall rationale and place in the EU, and the specific ways in which they interact with social town planning at the local and regional scales, are reviewed. The spatial policy framework is then outlined, together with its links with urban policy, and its relationship with the social sector and its contribution to meeting social town planning objectives are assessed. Finally, the concluding section looks at some of the issues that are to be faced in the next few years, particularly in the context of enlargement, reviews of funding programmes and the advent of monetary union (EMU).

One further preliminary point of clarification needs to be made. For convenience, terminology used to refer to European institutions is that established in the Treaty on European Union (the Maastricht Treaty) and in current usage at the time of writing, even when referring to earlier actions. The term EU is used for convenience throughout, since this is the term used in daily reference to the jurisdiction under consideration. This term has only been applicable since 1993, when the Treaty on European Union entered into force, and strictly speaking refers to the Union of three pillars established by that treaty, namely (1) the European Communities (the European Community, Euratom and the European Coal and Steel Community), (2) the Common Foreign and Security Policy, and (3) cooperation in the fields of justice and home affairs.

The social agenda under consideration here falls primarily within the first pillar, for which the EU has greatest independence of jurisdiction, and is seen as being an essential feature of the European Community (the word 'Economic' was taken out of its original title by the Treaty on European Union). Social issues may be expected to feature within the second and third, for which action is still largely intergovernmental, but as yet they do so only in the context of migration and the security consequences of the Schengen Agreement. The EU's institutional and decision-making structures as they relate to spatial planning are explained in R.H. Williams (1996), while Nugent (1991) provides a more detailed overall explanation.

The place of social policy in the EU

The first question to be asked is, why should an economic community concern itself with social matters? The key to understanding this connection lies in appreciating the sense attached to the word 'social' in the context of the social democratic consensus of the authors of the Treaty of Rome (1957). The word is used in the sense of 'social partners', i.e. the partnership between the two sides of the wealth-creation process: capital and labour, or entrepreneur and workforce. The underlying doctrine is that a true partnership is essential if the goals of economic prosperity, social cohesion and an improving quality of life for all citizens are to be achieved. In this sense, therefore, social policy is an employment policy not a welfare policy. This is why the the UK minister who attends the Social Affairs Council of Ministers is the Secretary of State for Education and Employment, or a junior minister responsible for employment.

As originally conceived, social measures are those which support the two sides of the partnership and their different needs, so that they are able to continue in partnership rather than in conflict with each other. In the years since the original treaty, the scope and reach of EU social policy has been extended in response to the changing political agenda and in association with other EU policies and programmes. Not least of these, in the context of social town planning, are those initiatives directed at different social groups within urban communities.

Terminology

The use of the English language to express non-British concepts in EU policy-making has been termed 'Euroenglish' (R.H. Williams, 1996, pp. 55–62). The distinct sense attaching to 'social', explained above, may be considered an example of this. Given the overall focus on social town planning, other terminological issues need to be discussed in respect of the rest of this expression.

The term 'town planning', and its legislative equivalent 'town and country planning', are well understood as generic terms in the British context. However, they connote a distinctly British concept of the scope and purpose of planning, which is seen by many in other EU member states as being very narrowly focused on urban land-use issues, controls and physical design. It translates as *urbanisme*, *Städtebau*, etc. 'Spatial planning' is the generic term that has come into use in the EU context, and increasingly now in the domestic context as well. It is a more all-embracing term for policy-making in respect of all spatial scales, from local urban and rural planning to the regional, national and supranational scales. It is the Euroenglish translation of *aménagement du territoire* (which in French usage has a more economic emphasis), *Raumplanung* or *ruimtelijke ordening* (which can be physical, or can be spatial coordination of policy sectors, including social).

In a sense, it follows from the above that this chapter should be about social spatial planning. However, this sounds like an oxymoron, so it is proposed to use the term 'spatial planning' in the context of EU spatial policy and planning initiatives, and the term 'social town planning' specifically to turn the focus of attention to the overall theme of the book.

European social policy

The main strands of EU social policy that are relevant in respect of the theme of social town planning and its EU context are discussed thematically. It must be recognised that the elements of EU social policy that are of significance to the theme of social town planning, however broadly defined, form only a small part of the whole body of social policy. It is helpful, therefore, to set out a brief chronology of its overall development before looking specifically at the elements of greatest interest, so that it is possible to see how the parts fit the whole. Reports are published by the Commission (CEC, 1996a, 1997a), and fuller discussion of EU social policy as a whole may be found in George (1991) and Moxon-Browne (1993).

Chronology of European social policy

Social policy dates back to the foundation of the EEC in 1958. The Treaty of Rome, in Articles 117–27, contained the first references to a European social policy and created the legal basis for the European Social Fund (ESF), which came into being in 1960. The policy was directed at issues of free movement of workers and assistance for migrant workers. The scope of the ESF was periodically reviewed and extended. Reforms extending the scope of the ESF which came into effect in the early 1980s led to a convergence between the possibilities offered by the ESF and the range of measures being adopted as part of local authority urban economic development initiatives at that time. Growing awareness of the ESF on the part of British local planning authorities thus led to the utilisation of the ESF to support many local economic development initiatives.

The ESF then operated separately from the other structural funds, but in 1988 regulations were adopted for the coordination of the three structural funds, i.e. the ESF, the European Regional Development Fund (ERDF) and the guidance element of the European Agricultural Guidance and Guarantee Fund (EAGGF) fund (R.H. Williams, 1996, pp. 82–3). From 1989, they have operated in respect of the same set of objectives. Some of the objectives are defined in spatial terms, others in social terms (see Box 9.1). Nevertheless, the structural funds remain the responsibilities of different Directorates-General of the Commission: for the ESF, DG V, the Employment, Industrial Relations and Social Affairs Funds; and for the ERDF, DG XVI, the Regional Policy and Cohesion Funds.

Meanwhile, the Single European Act of 1987 extended the scope of qualified majority voting in the social sphere and created a competence over the working environment and health and safety of workers. It also laid the foundations for the single market programme, leading to completion of the Single European Market (SEM) and extension of the four freedoms – free movement of capital, labour, goods and services – by the end of 1992.

The 1990s have seen the implementation of the Cohesion Fund, a facility introduced in the Maastricht Treaty to promote the economic and social cohesion of the EU as a whole by providing substantial financial assistance for infrastructure and environmental improvement in the poorest member-states. Funding is available to those member-states whose GDP (gross domestic product) per capita is below 90 per cent of the EU average. Currently, this applies to Greece, Spain, Portugal and Ireland, where it has supported quite substantial investment. The Maastricht Treaty also included the Social Chapter, discussed below.

Alongside the social agenda, the 1990s have also seen the emergence of the spatial agenda. Spatial policy has acquired a much higher profile, and is coming to the attention of town planners operating at the local and regional levels in a variety of ways. There are many aspects of the spatial policy agenda (see R.H. Williams, 1996, 1997, 1998). One that is of particular importance here because of the link it represents with the social agenda is that of the EU urban policy agenda. There were hopes that the Maastricht summit would agree to add an urban competence to the treaties, but this did not occur. Likewise, there was strong pressure on the

Amsterdam summit in 1997 to the same end. Although the Treaty of Amsterdam does not create an explicit urban competence, the overall EU urban agenda now has some momentum, as the Urban Forum is intended to demonstrate.

Following the Amsterdam summit, the EU's social agenda and all associated policies will be influenced by the next major steps towards European integration, namely the single currency and enlargement. An initial set of proposals concerning enlargement and the future of the structural funds was published in 1997 under the title Agenda 2000 (CEC, 1997b). Meanwhile, eleven member-states – all, apart from Denmark, Greece, Sweden and the UK – have signalled their intention to join the single currency in the first wave. Issues of employment and structural funding will be affected, but exactly in what ways and to what extent are still matters of speculation.

Structural and cohesion funds

Although the theme of this chapter suggests that, of the structural funds, the ESF is the one that should be given most attention, many specific initiatives of particular significance to social town planning at the local scale fall within the framework of the ERDF, and the agricultural fund is also of significance in respect of rural planning.

Since their coordination in 1989, the structural funds have been disbursed in response to a set of overall objectives to which all specific operations must relate. Five objectives were set (in effect six, as the fifth had two parts). These were revised in 1994, and a sixth objective was created for low-density arctic areas as part of the 1995 enlargement negotiations (R.H. Williams, 1996, pp. 90, 120–4). The six objectives applicable for the period 1994–9 as set out in Box 9.1.

Box 9.1 **The objectives of the structural funds 1994–9**

Objective 1: economic adjustment of regions whose development is lagging behind

Objective 2: economic conversion of declining industrial areas

Objective 3: combating long-term unemployment and facilitating the interaction into working life of young people and of persons exposed to exclusion from the labour market

Objective 4: facilitating the adaptation of workers to industrial changes and to changes in production systems

Objective 5a: adjustment of production and marketing structures for agricultural and fisheries products

Objective 5b: economic diversification of rural areas

Objective 6: economic adjustment of regions with outstandingly low population density

Objectives 1, 2, 5b and 6 are explicitly spatial, in the sense that their territorial application is precisely defined. It matters greatly to the local planning authorities concerned whether or not they fall within an objective area. Several of the specific measures of a social town planning nature are available only within these areas. Objectives 3 and 4 are so-called horizontal measures, applying throughout the EU. These have essentially social objectives. The ESF funds these and also contributes to activities under the spatial objectives. Objective 4 funding was not taken up by the UK under the Conservative government for ideological reasons, however, so the UK share of the structural funds was less than it might have been during this period. After the 1997 change of government, the Commission agreed a programme for objective 4 in the UK, with a budget of 224 million ecu, for the period 1997–9.

Within each region or member-state (depending on complexity and scale), a programme of activities and priorities for expenditure is agreed between the Commission and the member-state responsible. This may be in two parts, a Community Support Framework and a set of Operational Programmes, or in the more recent or more straightforward cases, a Single Programming Document. The latter is the usual arrangement in the UK, with one for each region.

The present basis for disbursing the structural funds runs until the end of 1999. Negotiation is under way for the next phase, 2000–6, based on the Agenda 2000 discussion document (CEC, 1997b). This proposes a simplified structure with three broad objectives (see Box 9.2). Extensive negotiation will take place, possibly extending into 2000, before agreement is achieved. This will take place with an eye fixed on the likely extent and impact of the next enlargement. Agenda 2000 is as much concerned with enlargement as with the structural funds because the two issues are closely linked. Structural funding arrangements for new member-states from central and eastern Europe will inevitably mean a reduction in the funding available for conventional projects in regions that have hitherto relied extensively on the funds, not least many parts of the UK. An additional factor, not addressed explicitly in Agenda 2000, is the advent of EMU and the role that the structural funds may play in compensating the economically weak regions of participating member-states.

Box 9.2 **Objectives proposed in Agenda 2000**

Objective 1: economic adjustment of regions whose development is lagging behind

Objective 2: economic and social restructuring of regions suffering from structural problems

Objective 3: development of human resources

NB: definitions and wording are subject to revision.

Social issues are prominent in Agenda 2000. The main role of the ESF will be to combat social exclusion, supported by the employment title in the new treaty. The proposed new Objective 1 is broadly equivalent to the present, but may incorporate some aspects of Objective 6. It will account for about two-thirds of structural fund expenditure, directed at regions with a per capita GDP of no more than 75 per cent of the EU average. The poorest UK regions hover around this threshold (the Highlands and Islands region, Northern Ireland and Merseyside are currently Objective 1 areas), but it may be that the first two of these will cease to be eligible after 1999 under the new criteria, when Merseyside may be joined by South Yorkshire under the new Objective 1.

The new Objective 2 would be directed at regions suffering structural problems of urban unemployment or rural depopulation through economic change. Priorities will include problems arising from loss of traditional industries, especially fisheries, youth unemployment, education, training in new technologies, environmental protection and combating social exclusion, especially in urban areas. This objective would apply to regions with unemployment above the EU average, high youth unemployment and problems of social exclusion in major cities. Some existing Objective 2 or 5b areas will be eligible, along with some areas currently in Objective 1, but there will certainly be several parts of the UK currently designated under one of the spatial objectives that will lose their status, because the UK enjoys relatively low official unemployment rates, not least because of successive redefinitions. For such areas, transitional arrangements for a phased withdrawal of funding over four years are proposed.

Objective 3 is explicitly social, and was proposed in anticipation of the inclusion of the new Title VIII: Employment, in the Treaty of Amsterdam (House of Commons, 1997a, pp. 29–31). It will apply throughout the EU, giving priority to access to employment, lifelong learning and local employment initiatives. It will be used as part of an EU employment strategy, with four themes: economic and social change; education and training systems; labour market policies; and combating social exclusion (CEC, 1997b, p. 24, 1998).

The ESF and social town planning

It would be impossibly tedious to summarise the wide variety of ways in which local planning authorities have utilised the ESF to support their urban policies. An overview of the ways in which EU programmes are integrated into the work of local planning authorities is found in DETR (1998b).

Much of the budget is used to support educational and training programmes. Some goes directly to universities and colleges, not least to support students of town planning, and much more is administered by the Training and Enterprise Councils (TECs). The local planning authorities may be working in partnership with the TECs, for example in urban regeneration and local economic development projects, to ensure that expenditure on training matches the needs of the local population and/or that of prospective incoming employers.

The Regional Fund and Community Initiatives

Discussion of the European Regional Development Fund (ERDF) as such falls outside the scope of this chapter (but see R.H. Williams, 1996; DETR, 1998b). Nevertheless, it plays a central role in EU funding for eligible local authorities, i.e. those falling within the spatial objectives 1, 2, 5b (see Box 9.1). It supports capital expenditure, unlike the ESF, and is very often used to support those developments, such as new industrial estates and infrastructure, for which the ESF is supporting training needs. For example, when the Tyneside Metro system was under construction, the ERDF supported the infrastructure costs while the ESF supported the training of drivers.

One use of the word 'social' in the ERDF context is worth noting, namely the concept of social infrastructure. Nowadays, it is normal for any city seeking to promote its image to emphasise its cultural attractions such as concert halls and art galleries. However, the traditional emphasis in regional policy in the UK has been on industrial development and the infrastructure required to attract and sustain it. It was the ERDF that alerted some local planning authorities to the possibility of support for social infrastructure projects such as concert halls and galleries, with the justification that enhancement of the quality of life supports economic development by increasing the attractiveness of a city to those making investment decisions or those seeking to attract key employees.

Around 10 per cent of the structural fund budget is allocated to Community Initiatives. These are programmes designed to address problems which occur in specific locations or types of area. There are thirteen Community Initiatives included in the 1994–9 agreement on the structural funds, resulting in 400 programmes of action. Several of the agreed priorities can be related to concerns of social town planning, but one, namely 'development of crisis-hit urban areas', is central to social town planning. The EMPLOYMENT and URBAN Community Initiatives address this priority, while others such as ADAPT and, in rural areas, LEADER, focus on economic development and skill-training needs. URBAN is discussed below, in the context of EU urban policy. The other Community Initatives that can support social town planning objectives are briefly outlined first.

A group of programmes under the title EMPLOYMENT seeks to counteract social exclusion by overcoming problems of access to the local labour market. This group has had three distinct strands since 1994: EMPLOYMENT-NOW, EMPLOYMENT-HORIZON and EMPLOYMENT-YOUTHSTART, plus a fourth, EMPLOYMENT-INTEGRA, approved in 1996. It brings together programmes dating from 1990 in the case of NOW and HORIZON, together with YOUTHSTART, which was new in 1994. NOW supports projects which seek to overcome the stereotyping of occupations, promote equal opportunities for groups under-represented in the local labour market and contribute to greater representation of women at higher technical and managerial levels. It can support the development of training and counselling materials and the creation of small businesses run by women, or help with costs of childcare and other responsibilities for dependents so that carers can participate in training.

HORIZON is aimed at overcoming the marginalisation and exclusion from the labour market of people with disabilities such as blindness or physical handicap, through support for work placements and placement services, training, counselling and the adaptation of tools and equipment.

YOUTHSTART promotes access to work or recognised education and training for young people under the age of twenty. It seeks to strengthen the links between education and employment, and reduce the numbers of school-leavers lacking qualifications.

INTEGRA aims to improve the employment prospects of disadvantaged groups who face particular difficulties in finding employment, and who often may be concentrated in certain neighbourhoods, contributing to a sense of social exclusion of these districts and therefore loss of cohesion of the city as a whole. Groups targeted include the long-term unemployed, refugees, ethnic groups and migrants with inadequate knowledge of the local language, single parents, the homeless, ex-offenders and drug addicts.

Urban policy

Although around 80 per cent of the EU's population live in urban areas, there has not hitherto been an EU urban policy, as such. It is nevertheless possible to point to several activities which, taken together, can be regarded as elements of an embrionic urban policy. Urban initiatives have come from both the environment and regional policy directorates of the Commission (R.H. Williams, 1996, pp. 204–17), particularly since 1990. Two of these are important in the context of social town planning: Urban Pilot Projects and the URBAN Community Initiative.

The title 'Urban Pilot Projects' is given to a series of demonstration projects designed to test new ideas for urban policy instruments, and to disseminate ideas for tackling urban problems. The first phase was in the early 1990s, with a second phase from 1995 to 1999. Great interest was generated, and thirty-two projects went ahead instead of the twenty-one originally intended by the Commission. Selection was on the basis of four principles:

- projects should address an urban planning or regeneration theme of European interest;
- projects must be innovatory and explore new approaches;
- projects should have clear demonstration potential so that lessons could be transferred to other cities;
- projects should contribute to the development of the region in which the city is located.

As always, the criterion of a balanced distribution between member-states and their major sub-national divisions was unstated but inescapable. Projects selected were based around four themes:

- economic development projects in areas with social problems, such as peripheral and inner-city residential areas with high unemployment and low levels of skills and access to training;
- areas where environmental actions can be linked to economic objectives;
- projects for revitalisation of historic centres, to restore commercial life to areas where the urban fabric has been allowed to decay;
- exploitation of the technological assets of cities.

Within the UK, projects in London and Paisley came under the first theme, and in Belfast and Stoke under the second, while Dublin and Cork had projects under the third theme. Huddersfield was successful in the second phase with its Creative Town Initiative, promoting the media and music industries through information technology.

The URBAN programme was launched in 1994, before the pilot projects had finished or been evaluated, as part of the 1994–9 agreement on the structural funds. This is perhaps an indication of the strength of the political coalition between the Commission and city authorities supporting greater intervention at the urban scale. It supports schemes to overcome serious social problems in the inner areas of designated cities with populations of over 100,000, through economic revitalisation, job creation, renovation of infrastructure and environmental improvement. Individual projects normally run for four years, must have a demonstrative character and potential to offer lessons to other cities, and form part of a long-term strategy for urban regeneration.

A total of eighty-five projects were approved by January 1997 (CEC, n.d.), of which twelve are in the UK and two in Ireland. Within the UK, URBAN projects in London (Park Royal and Hackney/Tower Hamlets), Birmingham, Manchester, Merseyside, Nottingham, Sheffield, Swansea, Glasgow, Paisley, Belfast and Derry were designated in the first round. Projects in Brighton, Bristol, Coventry and Leeds were added later in 1997.

One subtext for these initiatives was to support moves to add an urban competence to the EU treaties at the Amsterdam summit. Although this was not achieved, the new emphasis on tackling unemployment will have a strong urban dimension, and urban policy development is continuing within existing competences. In 1997, the Commission issued its White Paper *Towards an Urban Agenda* (CEC, 1997c). The next important step forward will be the establishment of the Urban Forum, to be inaugurated in Vienna in November 1998. Meanwhile, the UK presidency emphasised the need to coordinate the urban and spatial policy agendas (R.H. Williams, 1997) at the Glasgow Informal Council of spatial planning ministers in June 1998.

Agricultural Guidance

The Common Agricultural Policy (CAP) is often held responsible for excessive calls upon the EU budget and for generating surplus agricultural production. It should not be forgotten that in many ways the CAP is a social policy, supporting rural communities (Butler, 1993; Ilbury, 1998; Unwin, 1998). This feature is not particularly apparent in the UK, which has the most capital-intensive agriculture, but in many other parts of the EU, rural settlement patterns and population densities have been maintained with substantial support from the CAP. It would be very much a matter of concern to urban authorities if reduction in this support led to significance increases in migration to the cities.

The bulk of the budget goes to the guarantee section of the European Agricultural Guidance and Guarantee Fund (EAGGF). The guidance section operates within the framework of the structural funds and is the major source of funding for objective 5 (see Box 9.1). Many individual actions, especially in 5(b) areas, may be considered social town planning measures, albeit in a rural situation.

The Single European Market, single currency and Schengen

The foundations for the SEM were contained in the Single European Act (SEA), which entered into force in 1987. In spite of the name, this was a treaty that revised and added to the provisions of the earlier treaties. In respect of the social town planning agenda, as in many other respects, it did more to change the relationship between national governments and the EU than later and more controversial treaty amendments did.

The SEA laid the foundations for the single market by providing that all internal market legislation could be adopted by qualified majority voting, rather than unanimity. In other words, there was no longer a national veto on such proposals. It added competences to the EU of particular significance to social town planning through the insertion of Title V, Environment; and Title XVI, Economic and Social Cohesion; to the Treaty of Rome. This latter also created the legal basis for the ERDF.

The SEM is commonly associated with the date 1992, since the end of that year was the target set for its completion. The programme, as set out in the Commission's White Paper on the completion of the internal market (CEC, 1985), was based on the argument that, in order to maximise the benefits of the single market, the costs of non-Europe must be eliminated, i.e. a large number of tariff and non-tariff barriers needed to be removed (Cecchini, 1988). In the event, a total of 282 legislative measures were adopted. The key concept was that of the four freedoms: free movement of goods, services, capital and labour. These were extended to EFTA countries in 1993 by the agreement to create the European Economic Area. In 1998, this agreement covers eighteen countries, the EU plus Iceland, Norway and Liechtenstein.

The appeal of the single market programme to the UK government of the time lay in the aspects of freedom of trade, competition and access to European markets. However, its logic, following the doctrine of social partners, required that measures should be put in place to support people exercising their freedom to move to another member-state, whether or not they were moving to a job, seeking employment or wishing to live elsewhere non-employed. This implies transferability of social security and housing entitlements, so that migrants should enjoy the same rights as the locals. It also implies that it would be a distortion of competition, and therefore contrary to a fundamental principle of the EU, if one member-state sought to attract mobile firms and promote employment by exempting employers from social obligations to their employees applicable in other member-states. Hence the importance of EU social policy and the Social Chapter.

The single currency is the logical next step in the creation of one integrated common economic space. This will change many basic assumptions. One consequence of the single currency will be greater transparency in respect of all payments, whether as wages, social benefits, pensions, housing costs, etc., since they will all be expressed in the same currency units. People may be more tempted to move to gain economic benefit, whether from higher wages or lower housing costs. Meanwhile, trades unions will seek EU-wide agreements with firms or sectors of employment and will be less willing to accept different payment in different locations for the same work. It is not possible to do more than speculate what effects this greater transparency will have on pressures for development, but it will affect all sectors, commercial, industrial and residential. It will be necessary for local planning authorities to recognise this factor and anticipate its impact.

The Schengen Agreement dates back to 1985, but only became fully operational in 1997. It is incorporated in the EU treaties by the Treaty of Amsterdam. This provides for the elimination of all passport controls for travel between participating member-states. In the Treaty, the UK retained the right to maintain its border controls, as did Ireland and Denmark. Nevertheless, this agreement represents one more step towards integration. By removing impediments to movement in this way, local planning and housing authorities will increasingly need to anticipate pressures upon them generated by European, rather than national factors.

Social Chapter and Protocol

Although European social policy dates back to the Treaty of Rome (1957), the profile of this sector has increased since the Single European Act. One immediate effect was that legislation on the working environment and health and safety of workers could be adopted through qualified majority voting.

In 1989, a Charter of Fundamental Social Rights for Workers, known as the Social Charter, was adopted by eleven of the then twelve member-states. The UK did not wish to participate. This was a formal declaration that was not legally binding, but it was a significant declaration of intent, accompanied by an action plan of forty-seven proposals for legislation on various social policy issues. In fact, any directive adopted did apply to the UK since the legal basis was the Treaty of Rome, as amended by the Single European Act which the UK had signed. Several directives were so adopted, on issues such as working time and health and safety at work (UK Representation, 1997).

At the Maastricht summit it was proposed to insert into the treaty a new Social Chapter, based on the Social Charter. The UK was not prepared to agree to this, so the outcome was that a Social Protocol, to which all member-states agreed, was appended to the Treaty. The Protocol consisted of an agreement on social policy, with a text almost identical to that proposed for the Social Chapter, together with complex arrangements for a qualified majority voting system, allowing the other eleven member-states to adopt measures by qualified majority voting without the participation of the UK. These arrangements applied to the other fourteen member-states from 1995 to 1997 since, upon accession, Austria, Finland and Sweden also accepted the Social Chapter (R.H. Williams, 1996; p. 38). Following the change of government in 1997, the UK signed up to the Social Chapter, so the opt-out no longer applies.

The Social Chapter provides for measures to be adopted by qualified majority voting in the following areas:

- health and safety;
- working conditions;
- information and consultation of workers;
- equality between men and women at work with regard to labour market opportunities and treatment at work;
- integration of those excluded from the labour market.

A number of sensitive areas of social policy require unanimous agreement (i.e. member-states can exercise a veto):

- social security and protection of workers;
- protection of workers where their employment contract is terminated;
- representation and collective defence of the interests of workers and employees;
- conditions of employment for non-EU nationals;
- financial contributions for the promotion of employment and job-creation.

(UK Representation, 1997)

Most of these fall outside the normal scope of social town planning within a local planning authority context, but local economic development strategies often include policies aimed at the themes of the last item on each of the above lists. However, by late 1997 only two directives had been introduced under the Social Chapter, on works councils and on parental leave. Many UK companies have, in fact, chosen to introduce works councils in spite of the opt-out. Whenever possible, the Commission made proposals using as a legal basis the Single European Act, so as to minimise the impact of the UK opt-out.

Although the Social Chapter has been the subject of prolonged political attention in the UK (and of demonising by John Major's government), it represents only a small part of a much larger body of EU legislation concerning employment and social conditions. Many measures do not fall within the purview of social town planning, but a number do have implications for the economic development initiatives, especially those whose objective is to overcome social exclusion and promote participation in the workforce.

Equal pay for men and women has been required since 1975 under Directive 75/117/EEC. The principle of equal pay applies not only to wages and salaries but also to overtime and bonuses, sick pay and pensions. Equal treatment in statutory social security schemes was the subject of 1979 Directive 79/7/EEC, although retirement ages remained a matter for national governments. The UK is the only member-state with a five-year disparity between standard retirement ages (men retire at 65, women at 60), but this is to be equalised in 2010 following a judgement of the European Court of Justice concerning public sector workers.

The Commission has proposed a directive on the mobility and safe transport to work of workers with reduced mobility (CEC, 1992a), which would require provision of suitably adapted transport, and a report on possible supporting measures was presented to the Council by the Commission in 1993 (CEC, 1993a). However, this has not been adopted to date. Many planning authorities would see improving access to places of work for the disabled as an important social town planning objective, for which such a directive would provide support.

Promotion of homeworking and teleworking are seen as attractive possibilities by many planning authorities, especially in respect of communities not well served by public transport as well as for social reasons. They raise a number of social issues, particularly concerning the effects of working in isolation, working time and safety. No legislation is at present proposed, but the Commission is studying these issues and proposals may be put forward.

Spatial policy framework

This is not the place to undertake a full review of the EU's spatial policy framework, the emergence of which has been a noticable feature of EU policy development in the 1990s (R.H. Williams, 1996; pp. 218–27, 255–65, 1997, 1998). Nevertheless, it may be useful to draw attention to some key elements, and to some links between the spatial and the social sectors of EU policy, since social town planning, or spatial planning at the local scale, is subject to the influence of EU spatial policies as much as EU social policies.

Many sectors of EU policy have been essentially spatial in character, not least those aspects of the structural funds that most readily come to the attention of local planning authorities. A series of studies leading to the adoption of an overall framework for EU spatial policy have been undertaken during the 1990s (see, especially, CEC, 1991, 1994a), leading to the development of the European Spatial Development Perspective, a 'Complete Draft' of which was prepared under the UK presidency in 1998 (CSD, 1994, 1998). This is an attempt to provide a context for specific actions at the local and regional scale, both within and between member-states, at all the different spatial scales: supranational, transnational, inter-regional and cross-border and local (R.H. Williams, 1997, 1998).

Although the European Spatial Development Perspective (ESDP) will not be legally binding on member-states, it is hoped by the Commission that it, together with associated policy instruments, will help to achieve greater coherence between the different EU policies that have a spatial impact, and between EU policies and those of member-states.

A Community Instrument not hitherto mentioned, which will play a key role in this process, is INTERREG, which seeks to promote the development of spatial planning projects by local and regional authorities on a transnational partnership basis. The UK is a participant in three transnational planning regions under INTERREG IIC. At the time of writing, the Operational Programmes for the North Sea and the North-West Metropolitan Area have been approved, while that for the Atlantic Arc is still awaited. These programmes offer funding for projects, based on transnational partnerships. The first proposals for funding are being submitted during 1998.

One of the priority themes is for projects designed to promote balanced urban and regional development. Consequently, social town planning themes are likely to figure among projects supported by INTERREG, along with those on transport and environmental themes. Their essential feature is that all projects must have partners in different member-states and relate to the overal spatial planning concepts of the ESDP while directly addressing local needs, for instance by transferring the benefit of different cities' experience or sharing the benefit of expertise, to tackle urban problems at the local scale.

Conclusions

It has only been possible to give an outline in this chapter of the many ways in which EU policies and programmes may contribute to the achievement of social town planning objectives. Very often, programmes that local planning authorities undertake may not be identifiable explicitly as EU programmes. Urban Pilot Projects are an exception, but these are by their nature occasional opportunities for any authority.

The more normal situation is for an authority to use the appropriate category of structural funding to support programmes that are an integral part of its own strategy for tackling social town planning issues. The EU component may not be very visible, but it is often the case that strategies are developed in such a way that full advantage can be taken of EU funding possibilities. In order to do this successfully, the local authority needs to understand the European logic of the programmes, rather than rail against them for not appearing to be expressed exactly as a UK policy instrument might be.

In consequence of the lack of visibility, it is often the case that local planning practice, in respect of social town planning as in the economic and environmental contexts, is being influenced by the EU, giving rise to a degree of convergence between UK practice and that of other member-states (DETR, 1998b). There are a number of issues to look out for, as they could affect greatly the EU policy framework and funding opportunities that are open to the UK planning authorities after 1999.

First, increasing emphasis on participation within transnational and interregional planning frameworks is likely to be a prerequisite for funding. One motivation for participation in INTERREG is that it provides a learning experience in this respect. It is far from clear how significant the ESDP will prove to be, but ignoring it is unlikely to be the wise option, even for those responsible for social town planning at the most local scales.

Second, revision of the structural funds will take place with effect from January 2000, for the period to 2006. Agenda 2000 is merely the first step in a negotiation process that is likely to be long, arduous, and unlikely to be completed until the very last moment. Member-states and regions, especially those that risk losing designations and structural funding, are already staking out their positions and responding cautiously to the Commission's proposals.

Third, once ratified, the Treaty of Amsterdam will reinforce the EU's capacity to promote equality and fight discrimination and racism. The new Employment Title will introduce a new mandate to combat social exclusion through measures to ensure access for the whole labour market (CEC, 1998a; EIS, 1998).

Issues of the structural funds and social exclusion are complicated by the question of enlargement. The Treaty of Amsterdam paves the way for the next round of enlargement, with certain changes in the EU's institutional arrangements. A formal start to negotiations with eleven applicant countries took place in March 1998. Of these, six have been identified as suitable candidates for the first phase of the next enlargement: the Czech Republic, Estonia, Hungary, Poland, Slovenia and Cyprus. The others, Slovakia, Latvia, Lithuania, Romania and Bulgaria, form the second phase, while Turkey is still awaiting agreement to negotiate and Malta has withdrawn its application. All have per capita GDPs that are low by EU standards and will pose major problems of economic adjustment and environmental degradation.

In due course they will therefore make considerable demands on the structural and cohesion funds. Meanwhile, the relative position of existing member-states in respect of economic indicators will rise, and those that are at present at the prosperous end of the spectrum of those benefiting can expect to lose some of their present entitlement. The UK is in exactly this position.

Finally, and more speculatively, it could be argued that social integration within the EU has not proceeded as fast as economic and political integration. However, the potential impact of internal EU migration, following implementation of the Schengen agreement, enlargement and the single currency, may be considerable. If so, this would place great strains upon the social cohesion of receiving communities. Locations that may be sought out will include the most prosperous cities, and perhaps also those whose language is understood by the migrants. UK cities will be in the latter, if not the former, category. This issue has not been as widely recognised as that of migration from outside the EU, (but see King, 1998) although it is giving rise to tensions in some parts of eastern Germany. However, it may be that the planning impact of open borders, migrant workers and their access to housing and employment, will provide the big agenda for social town planning in the next decades.

10 Urban planning in Europe for health and sustainability

Colin Fudge

Introduction

Whilst chapter 9 detailed the main aspects of 'social planning' generated by the European Union, this chapter continues the European theme by putting emphasis particularly upon the 'health' dimensions (and the connections with sustainability and urban planning), making links both with contextualising global level organisations, especially WHO (World Health Organisation) and with the initiatives of nation states and individual European cities. This sets the scene for a more detailed, case-study-based investigation of the emerging links between health, housing and sustainability in chapter 11.

The European Union is one of the most urbanised continents in the world. The Union contains approximately 170 cities with more than 200,000 inhabitants and 32 cities with more than a million inhabitants. London and Paris are the only two metropolises with populations approaching 10 million. Over 80 per cent of the European population live in these towns and cities, making them the cultural, economic and innovative centres of Europe. They function as the generators of local, regional and national economies and together are the key localities in relation to European global competitiveness. They are also the centres of European social and cultural development and in recent times have undergone what is expressed as a 'renaissance' by some commentators. At the same time, many of these localities are confronted with serious problems: high unemployment, social and spatial segregation, social exclusion, concerns over their future economy, crime, the general quality of life, negative impacts on health and pressures on natural and historic assets. In addition, they are handling wider global and societal changes due to the globalisation of markets, shifts in demography and family structure, and new technological innovations.

Along with cities worldwide, European cities are facing up to the challenges that are reshaping their futures. In work carried out for the European Commission and the WHO over the last ten years, a number of significant and closely interrelated issues can be identified, which provide the agenda for policy development for both cities, health agencies, member-states, and the European Union. These include: the increased competition among cities and regions both within the Union and between the Union and the rest of the world; the accumulation of unemployment, poverty and social exclusion in the larger cities; issues concerning immigration; the increasing focus on sustainable urban development; the influence of changes to public expenditure and social insurance on cities; the increasing concern over urban health; the increasing inability to achieve access and mobility within and between cities; concerns over the quality of local democracy; and the requirements these challenges imply for urban management, urban leadership and governance.

Given this context, this chapter first explores an understanding of European urban change and the emerging European policies that have an urban dimension, culminating in a discussion of the Urban Communication *Towards an Urban Agenda in the European Union* (CEC, 1997c) and the Urban Action Plan. Alongside this an understanding of the nature of urban health is put forward. Both these areas of discussion lead to descriptions of two significant initiatives: the European Sustainable Cities Project and the WHO Healthy Cities Project. These, in turn, are brought together in a concluding section on the lessons to be learned from these initiatives for cities in the twenty-first century.

European urban change

Changes in the European urban system have been described in depth in a number of studies and reports. *Urbanisation and the Function of Cities in the European Community* (European Institute of Urban Affairs, 1992) still provides a stimulating understanding of urban change in Europe, as do the European Sustainable Cities Report (CEC, 1996c) and the Urban Communication *Towards an Urban Agenda in The European Union* (CEC, 1997c).

It is possible to identify a clear cycle of urban change in the European system during the post-war period, from urbanisation to suburbanisation, then de-urbanisation (also called counter-urbanisation) and, most recently, re-urbanisation, with close links between population shifts and changing economic fortunes. The largest industrial cities of the north and west experienced outward shifts in population and employment early on, while smaller towns and cities – especially those located in the south and west – grew. The period since the mid-1980s has witnessed a slowing of these population shifts in a period of economic recession and, most recently, a revival of population growth in some of the largest cities, linked in part to programmes of public and private investment in historic city centres. In the early 1990s the urban system was more demographically stable than in the period from the 1950s to the late 1980s, but cities are still vulnerable to change, especially from migration from eastern and central European countries. The Single Market and the enlargement of the European Union are further forces for urban change.

Towards an Urban Agenda in the European Union (CEC, 1997c) sees the main challenges facing European cities as revolving around the dynamics of urban change in Europe, owing to net immigration, national economic performance, structural change in the employment market coupled with the rapid growth and size of the service sector, the increasing importance of environment and quality of life conditions and the skills base and responsiveness of labour in locational decisions, and the expansion of the EU through the reunification of Germany and the accession of countries in central and eastern Europe (CEC, 1997c). Globalisation and the shift to the service sector has nevertheless not diminished the importance of space for economic development. There is, however, an emerging imbalance in the European urban system with the central gateway cities of Antwerp, Bremen, Rotterdam, Hanover, Lyon and Vienna, and medium cities located in the core of Europe, profiting more from European integration than cities on the periphery.

As these economic changes shape urban futures, inhabitants are increasingly concerned about the quality of their natural and physical environment and the quality of life their city provides. There is a growing dissatisfaction with the quality of air, water, the natural environment, safety, the quality of the built environment and the contribution of urban planning. The inhabitants' 'voice' in relation to these issues, although growing, is often fragmented, frustrated or unequal. There is also an imbalance in local democratic influence within and between cities across the Union. In terms of institutional response, cities across Europe are operating in very different legal, institutional and financial systems. Some local authorities, for example, operate within a greater tradition of local autonomy and wield larger spending powers than others.

A comprehensive review of the state of the built and natural environment in European cities is provided in chapter 10, on the urban environment, in *Europe's Environment: The Dobris Assessment* (EEA, 1995). This chapter in many ways complements the analysis of economic and social trends, for the links between urbanisation, economic change and environmental conditions are firmly established. Different patterns and stages of economic development generate different kinds of environmental problems and distribute them unequally, both within and between cities. In areas of both growth and decline the development and redevelopment of buildings and infrastructure have direct impacts upon natural ecosystems. Congestion, pollution from traffic, stress and noise have major consequences for health and, more generally, for the quality of life.

Europe's Environment analyses the quality of the physical environment in fifty-one European cities using data on twenty indicators, focusing on urban patterns (population, land use cover, areas of dereliction and urban renewal and urban mobility), urban flows (water consumption and waste, energy, transport of goods, waste production, treatment and disposal, and recycling) and urban environmental quality (air and water quality, noise, traffic safety, housing conditions, accessibility to green space and wildlife quality). This analysis has been updated (EEA, 1998) along with specific research developing an Urban Audit for Europe.

The European urban focus

Since 1991 the European Community, now the European Union, has sought consolidation of its actions for environmental protection, the reorientation of environment policy to promote the objectives of sustainable development in relation to towns and cities. These policy shifts have key implications for the urban environment. The principal developments are described and elaborated in *European Sustainable Cities* (CEC, 1996c). They include:

- The Treaty on European Union, Maastricht (CEC, 1992b);
- *Towards Sustainability*: the Fifth Environmental Action Programme (CEC, 1992c)
- European Spatial Development Perspective and Regional Policy (CEC, 1991);
- Committee on Spatial Development (CSD, 1994);
- Common Transport Policy (CEC, 1992a, 1992d, 1992e, 1993a, 1995a, 1995b, 1996d);
- Structural funds and their review (CEC, 1995c);
- Cohesion Policy (CEC, 1995d);
- Specific Directives e.g. on air quality (CEC, 1995e, 1995f);
- Delors White Paper (CEC, 1993b);
- European Environment Agency (EEA, 1995; CEC, 1994b);
- Agenda 2000 (CEC, 1997b);
- Urban Communication (CEC, 1997a; 1998).

In the field of environmental policy and, by implication, health, an integrated approach has been extensively pursued. It was first advocated in the Fourth Environmental Action Programme 1988–92. This led to the publication of the consultative Green Paper on the Urban Environment (CEC, 1990) and in 1991 to the Council of Ministers establishing the Expert Group on the Urban Environment.

The rationale for detailed consideration of the urban environment is set out in the Green Paper, which was a response to pressures from three sources: concern on the part of several European cities that a preoccupation with rural development within the European Commission was overshadowing the interests of urban areas, the commitment of the then Environment Commissioner, and a resolution from the European Parliament urging for more studies on the urban environment. The Green Paper is a significant milestone in thinking about the urban environment in Europe, principally because it advocated a holistic view of urban problems and a policy integration approach to their solution.

The Green Paper sparked a number of debates. The most heated, perhaps, concerned different views on urban form and the relationship between notions of compact cities and sustainable futures. While the urban form and the density of cities are clearly important, discussions since have widened the debate to consider ways in which cities and their hinterlands, regions and urban society are to be governed and managed to achieve sustainable futures. The Expert Group on the Urban Environment developed the European Sustainable Cities Project in 1993 (described in more detail later in the chapter), which led to a wider policy discussion, with an urban focus, in the European Commission.

In 1997 the European Commission published its Communication *Towards an*

Urban Agenda in the European Union (CEC, 1997c), which also established a process of consultation that culminated in the latter part of 1998 in a conference in Vienna at which the Urban Action Plan *Towards Urban Sustainable Development in the European Union* (CEC, 1998b) was discussed.

Further, the Fifth Framework for Research in the European Union, which will run from 1998 for four years, contains a strong focus on urban issues and includes a research area called the 'City of Tomorrow'. Fudge and Rowe (1997) in a report on the development of socio-economic environmental research for the European Commission suggest that the priorities for urban research to be included in the Fifth Framework Programme should include the following:

- upgrading current urban stock – which will also comprise the fabric of the 'City of Tomorrow' – towards sustainability goals;
- developing affordable and sustainable models for the future of access and mobility;
- reducing inequality and counteracting unemployment and social exclusion;
- investigating methods of implementing healthy public policy, including community safety;
- attuning urban economies to sustainability goals, at the appropriate scale and without exporting problems.

Thus the research agenda must include both the city and its hinterland; it must be underpinned by research into the changing nature of urban and social values; and it must examine the approaches to urban management and governance that are required for sustainable futures for cities.

Cities are increasingly at the centre of European policy thinking, even though the legal and constitutional competence is less clear. The 1998 Urban Forum in Vienna concludes a consultation period on the nature of 'urban sustainable development' in Europe and is intended to set out the emerging policy directions and processes for implementing the various strands that together might provide a European 'urban policy' that is also a European policy for 'urban sustainability'. Fundamental questions remain, including:

- What would a European 'urban policy' look like and how would it be implemented?
- How will social and health concerns be integrated with those of economy and environment?
- How will the accession of countries in central and eastern Europe affect this policy?
- How will a European 'urban policy' be designed to deal with global competition and improvements to the quality of life of all citizens whilst attempting to meet global environmental targets and contributions (e.g. the Kyoto Climate Treaty)?

Understanding urban health

Health is defined and accounted for in many different ways. The medical profession has dominated definitions and policy approaches in recent times, although the work of the WHO has always tended to broaden the definition from the medical focus on specific interventions to counteract individual disease. The WHO constitution definition is as follows:

> Health is a state of complete physical, mental and social wellbeing and not merely the absence of disease or infirmity. The enjoyment of the highest attainable standard of health is one of the fundamental rights of every human being, without distinction of race, religion, political belief or economic and social conditions.
>
> (WHO, 1994)

Health in this definition is not just about the absence of disease; it is about a 'state of being', a 'quality of life' that is influenced by individuals' age, sex and hereditary factors but also by their social, environmental and economic context. So their environment, their stress levels and their location matter and will have an influence on their 'quality of life' and their health. The early public health movement – and indeed the early days of the town planning profession – succeeded through considering not just the health of individuals but also the wider quality of the environment in campaigns, actions and utopian community designs that included a clean water supply, effective drainage and sewerage, housing reform, the provision of open space and the relationship between the different activities of home, work and leisure.

In the WHO document *The Solid Facts* (Wilkinson and Marmot, 1998) ten different but interrelated aspects of the social determinants of health are explained. These include:

- the need for policies to prevent people from falling into long-term disadvantage;
- how the social and psychological environment affects health;
- the importance of ensuring a good environment in early childhood;
- the impact of work on health;
- the problems of unemployment and job insecurity;
- the role of friendship and social cohesion;
- the dangers of social exclusion;
- the effects of alcohol and other drugs;
- the need to ensure access to supplies of healthy food for everyone;
- the need for healthier transport systems.

Each one of these aspects is examined through recent research evidence before policy recommendations are made and key references suggested. Taken together, 'the solid facts' provide the necessary understanding for the policies and actions that are needed to improve standards of health in the industrial countries of

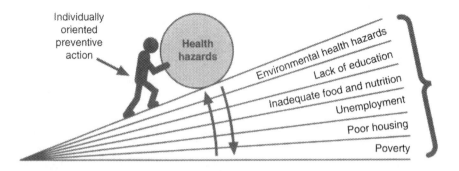

Figure 10.1 The health gradient.
Source: Adapted from WHO (1990), p. 13.

Europe. They provide information on the social and economic environment that would be conducive to higher standards of health.

Responding to these issues and making action plans for health requires policy influence in other sectors than that of health: for example, through town planning, economic development, transport planning and Agenda 21 processes. It also demonstrates that there is a significantly different range of opportunities and access to 'good health', dependent on one's position in society. In closing this section it is worth amplifying this point, as it should influence policy and planning thinking across the board.

> People's social and economic circumstances strongly affect their health throughout life, so health policy must be linked to the social and economic determinants of health.
>
> (Wilkinson and Marmot, 1998, p.8)

'Good health' involves reducing levels of educational failure, levels of job insecurity, and the scale and nature of income differentials in society. Policies for education, employment, town planning and housing therefore affect health standards. This can be best represented by Figure 10.1. The 1990 Green Paper on the Urban Environment which brought together quality of life concerns in urban areas of Europe made the link between good health, quality of life and the environmental and socio-economic context explicit in policy terms. This policy milestone introduced the Expert Group on the Urban Environment and the European Sustainable Cities Project.

The European Sustainable Cities Project

The European Sustainable Cities Project, supported by the European Commission, links policies for sustainable cities prepared by the Expert Group on the Urban Environment with implementation via the European Sustainable Cities

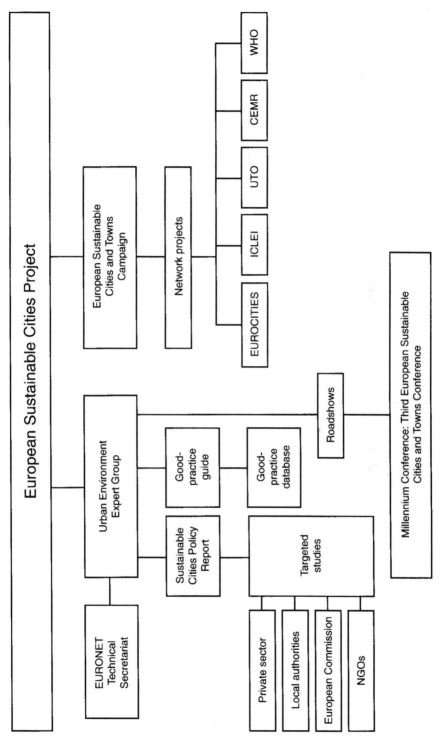

Figure 10.2 EU Expert Group on the Urban Environment: Sustainable Cities Project.

and Towns Campaign. These interrelationships between the policy work of the Expert Group on the Urban Environment and the Campaign, set in the wider context of the European Union's action in the field of urban environment, are focused around Local Agenda 21 (see Figure 10.2).

Local authorities and communities throughout Europe are now in the process of evolving their Local Agenda 21 strategies and a good deal of guidance has emerged on approaches and content. The UK is at the forefront of these initiatives and the implementation strategy provided by the UK Local Government Management Board offers a framework for action on a national scale, as does the guide prepared by the International Council for Local Environmental Initiatives (ICLEI) for the European Scale (LGMB, 1994, ICLEI, 1995).

The Expert Group on the Urban Environment, established by the European Commission in 1991 following the publication of the Green Paper on the Urban Environment, launched the Sustainable Cities Project in 1993, with the following principal aims:

- to contribute to the development of thinking about sustainability in European urban settings;
- to disseminate best practice about sustainability at local level;
- in the longer term, to formulate recommendations to influence European Union, member-state, regional and local levels of government.

The Sustainable Cities Project is based upon a major experiment in European networking which links over forty urban environment experts from fifteen member-states. The network of experts is supported by EURONET, the European research network based at the University of the West of England, which provides research expertise to the members of the Expert Group via its research and consultancy partners in Europe. A further thirty experts are involved in the Expert Group, including representatives from relevant Directorates-General of the European Commission and a range of international organisations with an interest in urban issues, including the Council of Europe, the Council of European Municipalities and Regions (CEMR), Eurocities, the European Academy for the Urban Environment, the European Foundation for Improvement of Living and Working Conditions, ICLEI, the OECD and the WHO. Current activities within the framework of the European Sustainable Cities Project include the following:

- *European Sustainable Cities* policy report (CEC, 1996c);
- Targeted summary reports;
- 'Local Sustainability the European Good Practice Information System' (http://cities21.com/europractice);
- Travelling dissemination conferences;
- European Sustainable Cities and Towns conferences;
- Network activities in support of the Campaign;
- Responses to the Urban Communication *Towards an Urban Agenda in the European Union* (CEC, 1997c).

The policy report

The preparation of the final *European Sustainable Cities Report* for the Lisbon Conference on Urban Sustainability in October 1996 represents one part of the developing agenda for urban sustainability. *European Sustainable Cities* (CEC, 1996c) explores the prospects for sustainability in urban settlements of different scales, from urban regions to small towns. However, the main focus is on cities, in line with the EU Green Paper on the Urban Environment. The policy report identifies the challenge of urban sustainability to:

> solve both the problems experienced within cities and the problems caused by cities, recognising that cities themselves provide many potential solutions. City managers must seek to meet the social and economic needs of urban residents while respecting local, regional and global natural systems, solving problems locally where possible, rather than shifting them to other spatial locations or passing them on to the future.
>
> (CEC, 1996c)

Sustainable development is identified by the Expert Group as a much broader concept than environmental protection. It has economic, social, health as well as environmental dimensions, and embraces notions of equity between people in the present and between generations. It implies that further development should only take place as long as it is within the carrying capacity of natural and social systems.

An important argument derived from these principles is that sustainable development must be planned for and that market forces alone cannot achieve the integration of environmental, social and economic concerns. The report seeks to provide a framework within which innovative approaches to the planning of sustainability can be explored. In this respect the report aims to identify a set of ecological, social, economic and organisational principles and tools for urban management, which may be applied in a variety of urban settings and which can be used selectively as cities move towards sustainability.

The report is aimed at a wide audience, for whilst elected representatives in cities, city managers/administrators and urban environment professionals have key roles to play in urban management for sustainability, successful progress depends upon the active involvement of local communities and the creation of partnerships with the private and voluntary sectors within the context of strong and supportive government framework at all levels. A core concern of the European Sustainable Cities Project.is thus the widest dissemination of the key messages of the work at both political and technical levels, raising awareness and developing new skills of sustainable urban management. EURONET and the International Council for Local Environmental Initiatives (ICLEI), have developed an information system entitled 'Local Sustainability – European Good Practice Information System' which addresses the above concerns in disseminating the work of the European Sustainable Cities Project and supporting local action for urban sustainability.

'Local Sustainability – the European Good Practice Information System' harnesses new information technologies to offer enhanced opportunities for integration of the work of the Sustainable Cities Project, linking policy to implementation, providing a 'hallmark' standard for good practice across Europe and ensuring the widest possible dissemination of this work. A system for researching and describing cases of good environmental practice by local authorities is being developed and evaluated through this project. The system incorporates written information (case studies) including illustrations, which are being made available through printed leaflets and provision on a World Wide Web site for on-line access. This system is to be developed further by the European Commission and via an interactive web site scheduled to be launched in November 1998.

The European Sustainable Cities and Towns Campaign

The European Sustainable Cities and Towns Campaign was initiated by eighty European local authorities through signing the Charter of European Cities and Towns Towards Sustainability, namely the Aalborg Charter, in 1994 (ESCTC, 1994). It was launched by the EU Environment Commissioner at the first European Conference on Sustainable Cities and Towns in Aalborg, Denmark on 27 May 1994. Any local authority (city, town or network of local authorities from any part of Europe) may join the Campaign by adopting and signing the Charter. The Campaign was strengthened further through the Lisbon Action Plan in 1998. There are now over 400 local authorities actively involved in the Campaign, from over 35 countries, with growing representation from central and eastern Europe.

The European Sustainable Cities and Towns Campaign is supported by major European networks and associations of local authorities. These organisations include the Environment Committee of the Council of European Municipalities and Regions (CEMR), the Environment Committee of EUROCITIES, the International Council for Local Environmental Initiatives (ICLEI), the United Towns Organisation (UTO) and the WHO. The supporting networks and associations have officially signed the Charter of European Cities and Towns Towards Sustainability (the Aalborg Charter) and thereby committed themselves to the principles and goals laid down in the Charter and the actions addressed in the Lisbon Action Plan, 1998.

They coordinate their efforts through a coordinating committee, working with their member local authorities in various ways to carry out projects supporting the Campaign. The coordinating committee supports the work of, and provides recommendations to the Campaign office. The European Commission (DG XI) and the city of Aalborg, both sponsors of the Campaign, also have seats on the coordinating committee, in addition to the chair of the Urban Environment Expert Group.

The objective of the Campaign is to encourage and support cities, towns, and counties in working towards sustainability and to promote development towards sustainability at the local level through Local Agenda 21 processes. This is to be achieved by strengthening partnership among all actors in the local community as well as through inter-authority cooperation, and relating this process to the European Union's action in the field of urban environment and the work of the Urban Environment Expert Group. It may be considered as the European Local Agenda 21 campaign.

The Campaign office's primary function is to facilitate the adoption of Local Agenda 21 processes, to promote sustainability objectives among the participating cities and to support the networks involved in the Campaign by providing general information and guidance, ensuring regular communication among the networks and between the networks and outside partners. This is achieved through the following:

- supporting the development of concrete projects of cooperation both within some of the participating networks and between cities;
- collecting and disseminating information on good examples at the local level, and making them available to Campaign participants through the good-practice guide of the Expert Group on the Urban Environment and Campaign newsletters which are provided on a regular basis in English, French and German;
- facilitating mutual support between European cities and towns in the design, development and implementation of policies towards sustainable development through workshops;
- following up EU initiatives relevant to sustainable urban development in order to inform local authorities of Commission initiatives;
- supporting local policy-makers in implementing appropriate recommendations and legislation from the European Union, through dissemination of information on European Union actions, policies and legislation in the Campaign newsletter;
- reviewing and further developing the Aalborg Charter.

Healthy cities

The WHO Healthy Cities Project is a long-term international development project that seeks to place health on the agenda of decision-makers in the cities of Europe and the rest of the world. It aims to build a strong lobby for public health at the local level, with a view to transforming policies and actions in cities to enhance the wellbeing of people who live and work in cities.

The project is one of WHO's main vehicles for implementing the strategy of Health for All (HFA). The concept of Health for All was adopted by the thirtieth World Health Assembly at Alma-Ata in 1977. The Ottawa Charter for Health Promotion further developed this thinking, advocating that action to promote health means building healthy public policy, creating supportive environments, strengthening community action, developing personal skills and transforming health services.

The idea of the healthy city, raised at the 'Beyond Health Care' conference in

Toronto 1984, is based on the notion that the city, or parts of the city, is the level of governance closest to the population and one where factors affecting health can best be influenced. Within the WHO definition a healthy city is not one that has achieved a particular level of health but rather one that is conscious of health in all its policy arenas and is striving to improve it. Between 1987 and 1992 WHO supported a pilot project for European cities to develop models of good practice in promoting health and positive urban health policy. In the first five years thirty-five project cities were participating in Europe, national networks developed in twenty countries and twelve groups of cities based on specific topics and interests (multicity action plans) were set up.

The first phase of the project between 1987 and 1992 emphasised advocacy and the development of structures and processes within cities to facilitate collaboration and partnership. In the second recently completed (see Box 10.1) 1993–8, the project focused on formulating and implementing urban policies targeting Health for All and on the development of city health plans with targets addressing issues such as sustainable development and equity. In this phase about 60 project cities have been designated and more than 550 cities are part of national networks in Europe.

Box 10.1 **Main requirements for participation in the second phase of the WHO Healthy Cities Project**

- All project cities should establish a widely representative intersectoral policy committee with strong links to the political decision-making system, to act as a focus for and to steer the project.
- All cities should appoint a person to be politically responsible for the project;
- All project cities should establish a visible project office that is accessible to the public, with a coordinator, full-time staff and an operating budget for administration and management.
- All project cities should develop a health-for-all policy based on the European Health for All targets and prepare and implement a city health plan that addresses equity, environmental, social and health issues, within two years after entering the second phase for old project cities, and within four years for new cities. Cities should secure the necessary resources to implement the policy.
- All project cities should establish mechanisms for ensuring accountability, including presentation to the city council of short annual city health reports that address Health for All priorities.
- All project cities should take active steps to take on the strategic action priorities of the WHO regional office for Europe and, in particular, to implement the European Tobacco Action Plan and the European Alcohol Action Plan.
- All project cities should establish mechanisms for public participation and strengthen health advocacy at city level by stimulating the visibility of and debate on public health issues and by working with the media.
- All project cities should carry out population health surveys, and impact analyses and, in particular, assess and address the needs of the most vulnerable and disadvantaged social groups.

(Adapted from Price and Tsouros, 1996)

A number of project cities, Copenhagen, Liverpool, Glasgow, have produced examples of comprehensive city health plans. The 1995 'Healthy and Ecological Cities' conference in Madrid produced a vast range of evidence of local initiatives, ranging from tower block energy conservation/anti-poverty projects in Sheffield to the restoration of a deprived area of Barcelona, from Cracow's pioneering Agenda 21 process to working towards 'child friendly' cities in Italy (Fudge, 1995; Price and Tsouros, 1996). At the Tenth Anniversary Conference in Athens in July 1998 WHO launched the third phase. This phase, which will last a further five years beyond 1998, is currently being developed by the WHO European office.

To illustrate further the Healthy City Project a number of examples have been selected and summarised. The case studies need to be followed in detail to obtain a full understanding of the processes at work and the politics of achieving a healthy city plan. Nevertheless, these 'potted' histories go some way to getting across the key points to the reader. They are taken from a significant WHO publication *City Planning for Health and Sustainable Development* (WHO, 1997).

Box 10.2 Glasgow: starting with a service-based plan

The city of Glasgow, Scotland (population 750,000) started the development of its city health plan (Glasgow Healthy City Project, 1995) with a model to be used as a plan for the agencies in the city. This was to be used to gain more support for collaborative work and to establish a firm foundation from which to develop a community perspective on planning for health in the city.

The process of developing the plan can be viewed as circular, beginning within and feeding back into the most powerful decision-making committees of the city.

The process started by obtaining commitment from the most important decision-making fora in the city. High-level commitment to the idea was necessary to provide access to senior management in the service agencies. The planning team conducted a series of meetings to develop an overview of the work of the agencies.

The City of Glasgow already collected an extensive range of health and social statistics, and these were fed into the process. The agencies themselves provided overviews of their work and how it related to a broad picture of health. A draft city document was produced over a year, based on these and personal interviews.

After extensive consultation, the Healthy City Project steering group, the local authority and health authority approved the draft. The draft was again circulated for updating and comment by all the service agencies.

This process of developing the plan within the agencies took about three years from start to launch, but it was only possible because of the strong collaborative structure and high-level commitment the project has worked to achieve and develop over its lifetime.

The launch of the document was not the end of the process but the beginning of the task of building wider community participation and development for the plan. After the plan was adopted and launched, a strategy for monitoring, review and development was put in place to monitor strategic targets and take forward the collaboration.

Box 10.3 **Bologna's health and environmental plan**

Bologna joined the WHO Healthy Cities Project in 1994, building on work taking place in the city since 1991. This work had developed the community capacity for health action, which led to the development of a self-protection health project and the opening of a health shop. This work involved the local community actively in gathering and producing information on health.

This collaboration at the local level is reflected in the activities and developments at the municipal level. The municipality of Bologna has signed the Aalborg Charter, joined the European Sustainable Cities and Towns Campaign and the WHO Healthy Cities Project and initiated a Local Agenda 21 process in 1996.

The Environmental Department is responsible for Local Agenda 21, coordinating collaboration both within the municipal board and at the local level. The commitment towards Local Agenda 21 and the Bologna Healthy City Project are clearly shown in the political plan for the city for the years 1995 to 1999, in which the main goal is the development of long-term action plans towards sustainability.

The objectives for the Healthy City include building strategic links with other sectors and organisations that substantially influence health and strengthening national alliances and support systems. The goal is to achieve a comprehensive city health profile and to plan to tackle such challenges as equity and sustainable development.

Because the principles and practice of Health for All and Local Agenda 21 overlap substantially, the muncipality intends to integrate environmental issues and health in developing a comprehensive municipal health and environmental plan. The Local Agenda 21 team and the Healthy City Project office are analysing the environmental factors in the city and preparing a city health profile.

Environmental indicators

The Agenda 21 team has already developed a set of eighty environmental indicators divided into three categories: status, pressure and response. The challenges identified are air and noise pollution, water resources, use of energy, waste disposal, transport, industrial activities, urban water planning and misuse of resources. This set of indicators will constitute a permanent database that will be updated regularly. From this, a sustainability index based on twelve indicators will be developed.

As the indicators were being developed, a series of public consultations and seminars were held in Bologna to develop a citizens' forum for Local Agenda 21. This forum, which has representatives from environmental groups, trade unions, the Artisans' Confederation, the Industrialists' Confederation, the professional associations, the Consumers' Association and other voluntary groups, will be involved in all significant stages of the preparation of the Local Agenda 21 plan. The citizens' forum will be the Local Agenda 21 team in developing and determining the twelve environmental indicators for the sustainability index. The tasks and priorities for the actual Local Agenda 21 plan will be discussed with the citizens' forum, and their suggestions and priorities will be taken into account. The final document will represent a collaborative effort between the community and the agencies and will provide a good base for working together in implementing the plans.

A city health profile and social equity indicators

Bologna's Healthy City Project completed the first city health profile in 1995 and is now working on selecting a set of indicators of social equity. This is seen as an important step in developing coordination of intersectoral action for health. The development of a set of indicators will help in the future definition of an index of social sustainability and capacity. This objective is still much more difficult to achieve for health and social challenges than it has been for environmental challenges.

The Healthy City Project team is developing methods and initiating discussions with the community about what challenges they perceive to be the most important. It is hoped that the health shop can be used as an important tool in finding and integrating the social and environmental factors affecting health in communities. A health network is being developed that will be decentralised in the nine districts of the city, enabling these communities to be reached in a new and more direct way. This network will have the health shop as a focal point.

The health and environment plan

The health and environmental plan will be based on the findings from the city health profile and the environmental reports and will serve as a guide for the priorities for action to be developed.

Box 10.4 Liverpool: developing a collaborative plan

The City of Liverpool felt strongly that developing health plans should not be seen as a one-off exercise that will set in motion all the work needed to solve the city's health and environmental problems at once. Instead a health plan should be viewed as a process of consultation, data-gathering and analysis, mobilising the resources required to make key improvements and open channels of communication between different sectors and communities in the city that can learn to work together on a continuing basis.

In Liverpool, the consultation with the service organisations and the community has been tightly coordinated with the planning process. The health plan is a 5-year strategy to improve the quality of life and health in the city. The plan has long-term and short-term targets and aims to get everyone moving in the same direction to take action on the underlying causes of ill health in Liverpool and to achieve the levels of health described in the document on health strategy for England.

Liverpool became one of the first cities to participate in the WHO Healthy Cities Project in 1988. During the first phase of funding, the Liverpool Healthy City Project was devoted to getting organised and to establishing commitment among all agencies to the philosophy, strategy and principles of Health for All. During the second phase of funding, the Project became involved in developing its own city health plan.

In 1993, new arrangements and structures were put in place for public health in the city, thereby bringing public health within the mainstream planning processes of the key authorities. The Healthy City Project was revised, with a brief to set up task groups on priority health challenges in the city. These task groups were made accountable to a newly formed Joint Public Health Team and to the Joint Consultative Committee. ▶

The role of the task groups was to develop strategic and operational plans towards the development of an overall city health plan. Task group topic areas included the national health promotion policy key topic areas of cancer, heart health, sexual health and accidents. A fifth task group was set up on housing for health to address the underpinning factors of ill health such as poverty. Housing was considered to be an extremely good indicator of poverty.

Members of the task groups included representatives from the health authority, the Liverpool City Council, the voluntary sector, the Trades Council, the City Health Promotion Department, community health councils, universities, the Chamber of Commerce and the communities of the city themselves. About 150 people were involved in developing the city health plan. The draft city health plan was launched for consultation in January 1995. The plan was produced in a number of versions and used in a number of settings with facilitators to ensure accessibility for all groups.

A video was also made to aid group discussion. Responses were elicited by questionnaires and letters.

The analysis of the consultation revealed that people widely supported the plan. City residents understood the links between the underlying causes of ill health and the main causes of premature death and illness and agreed that tackling these would have the greatest impact on health. The environment was ranked as the first of the top ten concerns expressed by local people. This included the environment in general, and specifically in relation to air pollution and the desire for a clean, green city. This was followed by poverty and unemployment, mental health, housing, access to health services, education and training, young people having more influence in the city, diet and nutrition, accidents and transport.

Comments from the consultation were then included in the revised city health plan, and the final version of the plan was launched on 7 April 1996 to coincide with World Health Day, which focused on healthy cities in 1996.

The city health plan reflects the need to address the wider socio-economic, environmental, political and discriminatory determinants of ill health. Action is being taken, for example, to tackle the problem of widespread poverty in the city. This action includes using new money such as European Union Objective 1 funding to target areas of the city with the intention of creating new work. Existing resources are being redirected towards city regeneration and investment in the economy, and the city is working towards balanced targets for environmental sustainability.

The plan is already influencing activity in the Liverpool City Council and Liverpool Health Authority. Some action has already been incorporated, and consequently the city is working and thinking differently. Gone are the old ways of thinking only about the present, short-term crisis management, exclusive ways of working and centralisation of planning.

The new ways of thinking and acting revolve around long-term neighbourhood planning, partnerships between agencies and local people, local action and cooperation, output that can be measured, concern for future generations and an increased understanding of health challenges and the impact on health of various policies and actions.

The impact of the plan on health will be measured. Indicators will be developed for measuring progress over the next five years of the plan, and a review of progress will be published annually.

(City of Liverpool 1996)

Box 10.5 Intersectoral planning for health in Belfast

Belfast started its process of developing a health plan with a very clear rationale and structure.

Since many different factors contribute to health, responsibility for its production rests with a wide range of individuals and agencies, each having a role to play individually and collectively in the production of good health in our city. Government has a strong role to play through its economic and social policies. Employers, large and small, have a role to play through their impact on the environment, the products they supply, the jobs they provide and the conditions under which their employees work.

Planning for health in the city therefore becomes the responsibility for all relevant sectors within the city, which impact upon it, environmental, social, education, housing and healthcare sectors. An aim and objectives for the city health plan were developed from this philosophy.

Aim

The aim of the city health plan is to create a vision for the health of the people of the city across all sectors and to develop integrated policies and strategies by which to achieve it.

Objectives

The objectives of the plan are:

- to develop a city health profile, including qualitative and quantitative data that will describe the health of the people of the city and the conditions in which they live;
- to make visible the challenges relating to health: for example, the current planned health activities and policies of statutory sectors and the facilitation of new integrated approaches to and coordination of all health and health-related activities within the city;
- to enable communities in the city to participate and influence decision-making processes about health and the provision of health-related public services;
- to provide a rational basis for decision-making, geared towards investing in health and reducing inequalities in health;
- to develop a monitoring and evaluation framework that will indicate the progress of strategies for action and measure the outcomes.

Process measures, outcome and time-scales are provided for each stage of the planning process. Overall and specific outcomes for the plan are also outlined. The overall outcomes desired by the city health plan are:

- a common direction towards health in the city for all agencies;
- improved coordination and synergy resulting from the process will enhance individual programmes.

The specific outcomes desired by the city health plan are:

- integrated policies evident in the future planning cycles of statutory organisation;

▶

- increased coordination between sectors on appropriate service delivery to meet community needs;
- evidence of change in the policy of statutory organisations that invest in good health:
- action strategies with time-scales and indicators that will lead to improvement in people's quality of life in such areas as environment and health, poverty and health, social networks for health and appropriate services;
- community involvement in future planning and development of policy and service in local areas;
- strengthened community participation in statutory organisations;
- ethnic minority communities being involved in assessing needs and developing action plans;
- an increased understanding of the wider dimensions of community needs and of the range of skills and resources within communities;
- new partnerships formed with the statutory, voluntary, community and private sectors at the community, city, country and European levels to sustain long-term action;
- the development of locality health plans;
- community needs being evident, both in policy changes and strategies for action;
- the development of support structures for negotiation;
- increased satisfaction from communities on delivery of services;
- evidence of collaboration between government departments that supports the long-term viability of communities;
- evidence that socially excluded groups are being involved in planning policy and programmes;
- evidence that organisations are developing and building sustainable policies and programmes in partnership with communities;
- evidence of capacity-building within communities.

Similar to all the city health plans developed within healthy city projects, the fundamental work of gaining commitment and having support and staff to carry out this work is part of the project.

(Belfast Healthy City Project 1996)

Conclusion

The quality of life in urban areas was raised as a major concern by WHO through its Healthy Cities Project launched in 1988 and by the European Community (as it then was) around the same time with its work leading up to the Green Paper on the Urban Environment. These two strands (now clearly interrelated) of policy innovation have been further enhanced by the UN series of conferences in the 1990s, particularly through the Rio Conference in 1992 and the Habitat II conference in Istanbul in 1996. These two conferences have provided vehicles for action at the local level in towns and cities across Europe – particularly through Local Agenda 21. In the concluding part of this chapter suggestions are provided on policy directions that may be followed within Europe that will affect urban planning into the next millennium.

Towards an Urban Agenda in the European Union (CEC, 1997c) initiated a debate conducted by the European Commission which culminated in the Urban Forum in Vienna in late 1998. The Urban Action Plan discussed at the Urban Forum included policy suggestions around the following interdependent themes:

- supporting urban productivity, employment and economic growth;
- promoting equality and social cohesion in urban areas;
- improving the urban environment and contributing positively to global sustainability;
- contributing to good urban governance and local employment.

Actions are based on five principles, namely: integration, subsidiarity, partnership, environmental efficiency and market efficiency. These stem largely from the work of the Expert Group on the Urban Environment's policy report, *European Sustainable Cities* (CEC 1996c). The more detailed suggestions argued for by both the Campaign and the Expert Group include changes to European policies, legislation, funding programmes and research to enhance urban sustainable futures and the quality of health. The policy suggestions that may influence urban planning for health and sustainable development at the local level include:

- As a prerequisite, the impact on towns and cities of existing EC policies, programmes and instruments should be evaluated.
- In looking forward to the Urban Forum and any emerging urban policy changes, the expansion of the EU should be included as an integral part of policy thinking.
- The contribution of towns' and cities' local policies and actions to global concerns such as climate should be made more explicit when changes to urban policies are being considered thus integrating both local and global concerns and making clear to European citizens the nature of their global responsibilities.
- It should be ensured that WHO Healthy Cities work is built into urban policy thinking in the European Commission.
- The notion of 'urban' in the Communication should be reframed to include the urban–rural interdependency and relationship. This emphasises the notions of ecological footprint and city region, both of which should be pursued via research programmes in DG XII, Science, Research and Development. In addition, further understanding is needed of the urban–rural relationship at the end of the twentieth century.
- There remains some concern that the 'natural environment' and biodiversity issues were apparently omitted from the Communication, even though they are important issues for urban areas. As well as policy changes, suggestions were made in relation to inclusion in revised urban pilot projects and research.
- There was considerable discussion of the nature of the urban economy and

the relative influence of global and local levels. A dual urban economy was postulated, made up of a global component that is difficult to control locally and a local economy that can be influenced. Pilot projects and research were put forward as suggestions for examining how the 'two economies' can be interrelated positively in urban areas.

- Some cities have been exploring the development of local food production as part of their sustainable urban policy. These cities are suggesting that European agriculture policy and funding needs to be changed to support such initiatives.
- All economic development and regeneration programmes funded by the Commission should be recast to ensure achievement of sustainable objectives and include sustainability criteria. In addition, some urban projects should be recast to combine urban and rural areas in joint proposals.
- Local authority procurement should be enhanced, together with the liberalisation of rules to achieve sustainable objectives. It was suggested that pilot projects should be set up, freeing some zones in Europe from procurement rules. These zones should be evaluated to see if sustainability objectives can be advanced through procurement.
- In addition, it was suggested that Europe-wide local authority procurement consortia be organised to improve purchasing's contribution to meeting sustainability objectives.
- Funding and policy should be reallocated to provide a strong focus on transport problems – the most common issue facing all towns and cities. DG VII, Transport, should take the lead in association with DG XI, Environment.
- The development of the draft ESDP (European Spatial Development Perspective), urban audit approaches and sustainability indicators should be continued, and the work of the European Environment Agency and the Commission combined to avoid duplication.
- Further work on 'urban management and governance' should be supported – this is critical at all levels of government and involves cultural change, institutional capacity and change, training, education and practice dissemination.
- Education and awareness programmes should be continued and expanded. They should be developed within a new across-the-Commission strategy focusing on urban sustainability. Existing approaches, such as the European Sustainable Cities and Towns Campaign, should be utilised and built on.
- As a matter of urgency, practice-sharing through a database, interactive web sites, partner forums, and conferences, seminars and workshops such as the annual Campaign day should be expanded and implemented.

The outcomes of the Urban Forum are likely to be influential on urban planning in Europe in the next millennium. It is essential that the 'voice' of cities and towns in Europe is heard and listened to in relation to the emerging urban policy directions in Europe. It is clear that much has been learnt in practice through Local Agenda 21 and Healthy Cities programmes in the

1990s; it remains to be seen how the principles of integration, subsidiarity, partnership, environmental efficiency and market efficiency stand up to the pressures from global competition and the diversity of local practice in European towns and cities.

11 Planning for health, sustainability and equity in Scotland

Janet Brand

The international planning agenda in Scotland

This chapter illustrates how the new international agenda on sustainability (stemming from the Rio Declaration and Agenda 21) may assist, at the local level, to resolve the social/spatial dualism inhibiting the fulfilment of social planning within the current UK statutory planning system, by using comparative examples of implementation through local partnership studies in Scotland. As Scotland prepares for a devolved parliament, it is evident that the processes, and their relationship to the planning system, are taking a form different from those in the rest of the UK, providing interesting comparisons with the English situation in respect of the conceptualisation of social town planning.

First, sustainability is defined in more detail than in chapter 1 and the role of Local Agenda 21 in delivering sustainability is explained. Then a series of case studies is provided in which multi-agency initiatives are discussed. They include a national example (the sustainable communities programme), a city example (the capital city of Edinburgh), an international rural example with Scottish projects (CADISPA: Conservation and Development in Sparsely Populated Areas), and an example of a remote rural island (Islay). Finally, some conclusions are offered relating to the challenges of pursuing the objectives of achieving a more sustainable future.

Sustainability revisited

We may think of sustainable development as delivering basic environmental, economic and social needs to all without threatening the viability of the systems upon which these needs depend. The Brundtland Report defines sustainable development as 'Development that meets the needs of the present without compromising the ability of future generations to meet their own needs' (WCED, 1987, p. 43). As will be explained, the concept of sustainability can fully incorporate, indeed requires, the two ingredients – health and equity – mentioned in the chapter title.

As discussed in chapter 10, the World Health Organisation (WHO) defines health as 'a state of complete physical, mental and social well-being and not merely the absence of disease and infirmity'. The definition recognises that health is the consequence of many different influences including social, environmental, economic and biological factors. Sustainability implies the removal, as far as is possible, of the causes of ill health, so that people can enjoy membership of, and make a more positive contribution to their local community (Brand, 1996).

> Sustainability is not a threat . . . if the sustainability message is to be successfully delivered to the wider public (and especially industry), it must have a positive message – sustainability is about improving the quality of people's lives, making business more resource efficient and so keeping costs down, etc. It is not about sitting in the dark and having nothing but porridge to eat.
>
> (FoE, n.d., p. 23)

A considerable literature has built up attempting to define and refine the meaning of sustainable development and address its key principles (see, for example, Barton *et al*,. 1995; Blowers, 1993; McLaren *et al*., 1998; Pearce, 1989; Selman, 1996; Walton *et al*., 1995) What becomes clear is that from small changes, maybe many small changes, in behaviour, grander and more far-reaching changes can grow. By the early 1990s, a consensual definition of the aim of sustainable development was:

> To promote development that enhances the natural and built environment in ways that are compatible with:
> 1   The requirement to conserve the *stock of natural assets*, wherever possible offsetting any unavoidable reduction by a compensating increase so that the total is left undiminished
> 2   The need to avoid damaging the *regenerative capacity* of the world's natural ecosystems
> 3   The need to achieve greater *social equality*
> 4   The avoidance of the imposition of *added costs or risks* on succeeding generations.
>
> (Blowers, 1993, p. 6; emphasis added)

Recognising that we do not yet have the institutional fabric – in governmental, political or administrative terms – to deal with sustainable development, Wellbank added the aspect of futurity or 'long-term effects' and offered four 'distinctive principles of sustainable development namely, long-term effects in intra-societal equity, inter-generational equity, (and) environmental sustainability' (Wellbank, 1994, pp. 15–16).

Much government policy and advice refers to sustainable development. In Box 11.1 we can see how summary definitions have evolved. In simple terms, we can identify three types of definitions, those which:

* talk of reconciling the need for economic development with environmental protection and basically view the concept as 'environment versus jobs';

Box 11.1 **Selected definitions of sustainable development produced by the UK government**

The Conservative Administration – up to May 1997

Sustainable Development: The UK Strategy
Sustainable development does not mean having less economic development: on the contrary, a healthy economy is better able to generate the resources to meet people's needs, and new investment and environmental improvement often go hand in hand. Nor does it mean that every aspect of the present environment should be preserved at all costs. What it requires is that decisions throughout society are taken with proper regard to their environmental impact.

(DoE, 1994a, p. 7)

Department of the Environment, Planning Policy Guidance
Sustainable development seeks to deliver the objective of achieving, now and in the future, economic development to secure higher living standards while protecting and enhancing the environment.

(DoE, 1997a, p. 3)

The Labour administration – from May 1997

Speech by the Minister for Agriculture, the Environment and Fisheries at the Scottish Office, Lord Sewel
For us, in Government, it (sustainable development) has three arms:
• concern for the environment
• economic growth
• social progress
. . . We are taking steps to integrate these issues which are inextricably linked in the quest for sustainable development.

(Sewel, 1997a, p. 4)

Deputy Prime Minister, the Rt. Hon. John Prescott, in the foreword to opportunities for change
Sustainable development is 'all about, a new and integrated way of thinking about choices right across Government, and throughout society, so that we can all share in the highest quality of life now, without passing on a poorer world to our children.

(DETR, 1998c, p. 1)

Opportunities for Change: Consultation Paper on a Revised UK Strategy for Sustainable Development
Sustainable development is a very simple idea. It is about ensuring a better quality of life for everyone, now and for generations to come. To achieve this, (it) is concerned with achieving economic growth, in the form of higher living standards, while protecting and where possible enhancing the environment . . . and making sure that these economic and environmental benefits are available to everyone, not just to a privileged few.

(DETR, 1998c, p. 4)

- look to balance the needs of the environment with development;
- incorporate social progress and equity.

What has become clear is that this is not just a physical concept but one which is holistic, involving a new way of thinking. Turning sustainable development into practice requires action in three related spheres of activity: the economic, the social, and the environment. Figure 11.1 offers a conceptual framework whereby, at the intersection of all three spheres of economic vitality, environmental integrity and community well-being, sustainable development will result; this 'requires innovation, demonstration and replication of best practice at every level of society and in every walk of life' (Forward Scotland, 1997, p. 9 cf. DETR, 1998c, pp. 4–5). Clearly this means involving 'everyone', including women and other so-called minority groups within the community. Principles to guide action are frequently expressed and, as demonstrated in Box 11.2, translated into practical guidelines for local sustainability. One means of achieving sustainability is through the Agenda 21 process (Sewel, 1997a, p. 5).

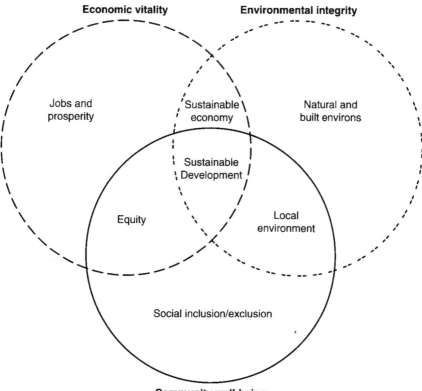

Figure 11.1 A model of sustainable development.
Source: Forward Scotland (1997), p. 9.

Local Agenda 21

A significant outcome of the Rio Earth Summit in 1992 was the adoption of Agenda 21 (UNCED, 1992; Keating, 1993). In the form of a comprehensive programme, it provides a framework for action into the twenty-first century, by governments, development agencies, UN organisations and independent sector groups in every area where human economic activity affects the environment. It is more than an environmental charter and emphasises the need for global and local partnership for sustainable development. Adopted by most of the governments at the Earth Summit, there is a strong moral obligation for them to ensure its implementation, although it is not a legally binding document. Agenda 21 is a lengthy document, with forty chapters split into four sections: many have reviewed its contents (see, for example, Blowers, 1993; Keating, 1993; McLaren *et al.*, 1998; DoE, 1994c; LGMB, 1993a).

In summary, 'emphasis is placed on social and economic dimensions, the need to combat poverty, strengthening the role of major groups (including women, local authorities, and NGOs (non-governmental organisations)) and the process of implementation, including partnership' (Brand, 1996, p. 60). In particular, Agenda 21 acknowledges poverty and health issues, and states: 'good health depends on social, economic and spiritual development and a healthy environment, including safe food and water' (Keating, 1993, p. 11) (reiterating WHO objectives).

Indeed, there are several pre-Rio international projects geared to similar objectives to those of sustainable development, and overlapping on Rio. For example, as explained in chapter 10 the Healthy Cities Project, formally launched by WHO in 1986, is an initiative designed to promote good health and prevent disease, and 'seen as a means of legitimising, nurturing and supporting the process of community empowerment. Using community participation as a method, it seeks to reduce inequalities, strengthen health gain and reduce morbidity and mortality' (Davies and Kelly, 1993, p. 3) – thus to do social town planning. Glasgow is the only Scottish city participating in the WHO Healthy Cities project. Commencing in 1988, and 'being framed within a community development approach', it provides a 'bottom-up' approach which appears close to the spirit of Agenda 21 (Brand, 1996, chapter 10 above).

Of the statements in Agenda 21, as adopted by national governments, over two-thirds can only be delivered with the active commitment and co-operation of local government and local communities. Chapter 28 of the document calls upon each local authority to undertake a consultative process with its citizens and develop a Local Agenda 21. So it is for local authorities, with their communities, to design and deliver. Within the context of their national strategies for sustainable development, central governments may help the local process. Given the complexity of Agenda 21 and the need for the commitment and co-operation of UK local governments for its fulfilment within the Local Agenda 21 process, various national bodies have responded by offering guidance to the more local level (see, for example, CVS, 1995; UK Steering Group, 1994; LGMB, 1993a, 1993b, 1994, 1996).

Throughout Agenda 21 there are calls for new approaches to decision-making, emphasising inter-sectoral co-operation, local consultation and greater democracy. Thus authorities and their communities need to develop structures through which they can work together towards a sustainable society. (The editor notes these are all approaches also favoured by the 'women and planning' movement.) Local Agenda 21 should not affect only planning and environmental health departments, it should be an integrated strategy for sustainable development which is understood and implemented by all departments and be incorporated into all aspects of local authority work. This may involve reviewing and modifying existing programmes and policies: however authorities may find that some of their current work is already in keeping with Agenda 21. Authorities are encouraged to share information, foster partnerships, mobilise international support for local programmes, and specifically to ensure that women and youth are represented in the decision-making process (see Box 11.2).

Box 11.2 Practical guidelines for local sustainability

The *public trust doctrine*, which places a duty on the state to hold environmental resources in trust for the benefit of the public.

The *precautionary principle* (erring on the side of caution), which holds that where there are threats of serious or irreversible damage, lack of full scientific certainty shall not be used as a reason for postponing cost-effective measures to prevent environmental degradation.

The *principle of inter-generational equity*, which requires that the needs of the present are met without compromising the ability of future generations to meet their own needs.

The *principle of intra-generational equity*, stating that all people currently alive have an equal right to benefit from the use of resources, both within and between countries.

The *subsidiarity principle*, which deems that decisions should be made by the communities affected or, on their behalf, by the authorities closest to them (though the 'appropriate level' of decision-making may be problematic to determine in respect of international environmental issues or transboundary pollution).

The *polluter pays principle*, which requires that the costs of environmental damage should be borne by those who cause them: this may include consideration of the damage occurring at each stage of the lifecycle of a project or product.

(Selman, 1996, pp. 14–15)

Developing the Local Agenda 21 process is not proving to be an easy task, as a progress review demonstrated (Tuxworth and Thomas, 1996), and much remains to be done. All local authorities were to produce a 'Local Agenda 21' by 1996, and the UK government claimed commitment to this process (DoE, 1994a, 1994b). UK local authority re-organisation was a significant pre-occupation in the mid-1990s. However, many authorities have environmental charters outlining their commitments in areas such as energy conservation, transport, land-use planning and waste management. Several have developed local sets of indicators of sustainable development.

Many are working on the social dimensions of sustainability through programmes to improve damp housing, access to education and primary health care; others are encouraging good environmental practice amongst business and supporting community business initiatives. The UK government now wants 'all authorities to adopt Local Agenda 21 plans by the year 2000' (DETR, 1998c, p. 16).

Implementation through partnership

International initiatives

What must be emphasised is that Local Agenda 21 is not a product but a process. Given the scope of the Agenda, no one body or group can deliver it, and thus partnership with a range of bodies is essential. In 1993 an international conference, 'Partnerships for Change', was held in Manchester, UK: the main impetus was to assist the objectives of the Earth Summit in progressing a global partnership towards achieving sustainable development (DoE, 1994b). The findings give a positive message about the universal validity of partnerships, arguing that it would be difficult to discover a development that is to be sustainable unless all threads of a community are represented; this concept is idealistically expressed, thus: 'poor as well as rich; young as well as old; women as well as men; labour as well as employers and business/industry as well as environmentalists' (DoE, 1994b, p. 6). Indeed, concern with global distribution of resources, with Third World poverty (and First World wealth) is a key component of Agenda 21. Table 11.1 lists some of the main documents and related partnership associations to achieve sustainability. Involvement by the UK government and NGOs work at the UN level and involvement of other global 'planning' bodies, such as WHO, are all essential.

European support

There are also many partnership opportunities with ongoing European Union initiatives (CEC, 1990, 1991; and cf. chapter 9 above). A main goal of the Fifth Environmental Action Programme (CEC, 1992c) is to raise awareness of environmentally friendly behaviour amongst three identified groups of actors: government, consumers and the public. The basic theme running through *Europe 2000+* (CEC, 1994c) relates to closer co-operation over territorial planning in Europe. A major objective of the European Union is to achieve sustained and non-inflationary growth respecting the environment . . . . The pursuit of this aim implies the integration of environmental considerations into other policies.' (CEC, 1994c, p. 17; see also Revie, 1997a). The environmentally blind expenditure decisions in the use of structural funds and in the development of Trans-European Networks have been much criticised, but such is the effect of compartmentalisation of policy directives within the EU. EU Directives, including those of Environmental Impact Assessment, Habitats, Social Policy and, as proposed, on Strategic Environmental Assessment, are binding on all member-states. Yet initiatives such as the European Spatial Development Perspective (ESDP) are seeking

Table 11.1 Sustainable development: a chronological list of key international and UK national policy documents

Date	Organisation	Policy document title
1980	International Union for Conservation of Nature and Natural Resources (IUCN)	*World Conservation Strategy*
1987	World Commission on Environment and Development (WCED)	Our Common Future (The Brundtland Report)
1990	UK government	*This Common Inheritance: Britain's Environmental Strategy* (plus subsequent annual reports)
1991	European Commission	Europe 2000
1992	United Nations	Rio Declaration on Environment and Development and Agenda 21
	European Commission	*Towards Sustainability: Fifth Environmental Action Programme*
1993	LGMB	A Framework for Local Sustainability
1994	UK Government	*Sustainable Development: The UK Strategy*
	European Commission	*Europe 2000+*
1995	UK government	The Environment Act
	The British Government Panel on Sustainable Development	*The British Government Panel on Sustainable Development: First Report*
	UK government: House of Lords	*Report from the Select Committee on Sustainable Development*
1997	Labour Party	Manifesto placing 'concern for environment at heart of policy-making.
	UK government	*This Common Inheritance:* 1997 report
	European Commission	*European Spatial Development Perspective*
1998	UK government	Opportunities for Change: Consultation *Paper on a Revised UK strategy for Sustainable Development*
	UK government	*Our Healthier Nation: A Contract for Health* (Green Paper)

to deal with this spatial blindness, particularly in respect of the need to plan for sustainability in relation to specific cities' and regions' needs in the EU, bringing environmental policies from the global and pan-European level 'down' to city specific implementation (Morphet, 1997, p. 266).

Policy areas suggested for action at regional and local urban level include co-ordinated land-use programmes, promotion of sustainable strategies and urban/rural partnerships. Developing the ESDP into a practical tool, with the emphasis on voluntary co-operation between member-states, will entail consideration of a number of thorny issues, not least of which will be the principle of subsidiarity. Many of the actions that might be envisaged as components of the

ESDP are already the preserve of central or local authorities and their relationship to the ESDP will be an important factor in its development (Healy, 1997, p. 13).

UK government policy

As a member of both the UN and the EU, the UK government has been progressing its own national policy stance towards the issue of sustainable development. Table 11.1 indicates the key policy documents, which, together with the key agencies, will now be summarised.

In 1990, the UK government published a strategy for the environment in the White Paper *This Common Inheritance* (DoE, 1990). Claiming support for the principle of sustainable development, it stated: 'The Government's approach begins with the recognition that it is mankind's duty to look after our world prudently and conscientiously . . . the Prime Minister [Thatcher] . . . reminded us that we do not hold a freehold on our world, but only a full repairing lease' (p. 10). The strategy included specific targets and objectives for policies in different areas; progress and new commitments have since been reported through annual reports which reiterate the importance of meeting the objectives of the Earth Summit and Agenda 21 (see, for example, House of Commons, 1997b; DETR, 1997b). Thus a process is emerging in the UK for updating policy and monitoring its implementation.

Produced in 1994 as a response to the UK government's commitment to meet the principles embodied in the Rio Declaration, *Sustainable Development: The UK Strategy* looks at challenges for the next twenty years. Setting a future agenda for, government, business, NGOs, and individual men and women, it stresses partnership approaches (DoE, 1994a). A UK Local Agenda 21 Steering Group was established and the Local Government Management Board took on the role of disseminating guidance throughout the UK (see, for example, LGMB, 1993a, 1993b, 1994, 1996)

Appointed in 1994 by the Prime Minister to advise on strategic issues arising from the Strategy and other UK post-Rio reports (DoE, 1994a), the first annual report of the Government Panel on Sustainable Development highlights issues requiring decisions or choices for the UK government and NGOs (DETR, 1997b). Further advisory bodies established in the mid-1990s include the UK Round Table on Sustainable Development (UK Round Table, 1996) and the House of Lords Select Committee on Sustainable Development (House of Lords, 1995). In addition, a number of quangos have statutory duties with respect to environmental development, for example the well-established Countryside Commission. In 1995 the Environment Act established Environmental Protection Agencies in the UK.

The election in May 1997 brought a change in administration following eighteen years of Conservative government. It has been argued that 'the pursuit of a sustainable future for our society has become a common goal across all parties' (Sewel, 1997b, p. 2). The new government's agenda includes a commitment to policies that cut across departments, social inclusion, welfare to work, sustainable development and best-values issues. Indeed, in a speech in June 1997, the Deputy

Prime Minister, John Prescott, pressed for local government to deliver on five priorities: integration, decentralisation, regeneration, partnership and sustainability. These are all inter-departmental and all need a local dimension to be implemented.

The Labour Party Manifesto claimed it would place 'concern for environment at the heart of policy-making, so that it is not an add-on extra, but informs the whole of government, from housing and energy policy through to global warming and international agreements' (Labour Party, 1997). Later that year, a minister at the Scottish Office interpreted this as 'a pledge to sustainable development itself . . . [making it] the fundamental driving principle of all our policies' (Sewel, 1997b, p. 3). To drive home the manifesto's message, and recognising the need for an educational and learning process at the heart of government, a new Cabinet Committee, known as ENV, was quickly established. Chaired by the Deputy Prime Minister, 'its main role is to achieve that coherence, that integrated approach to policy, demanded by sustainable development' (Sewel, 1997b, p. 3).

The previous government had appointed a number of Green Ministers (see House of Commons, 1997b back page) and this is maintained, in that the ENV committee has an associate group of Green Ministers 'charged with ensuring that the operations of the Government machine itself are compatible with our overall promise on the central place of the environment in policy-making' (Sewel, 1997b, p. 4). Furthermore, a new Parliamentary Environmental Audit Committee has been set up to monitor progress on the delivery of sustainable development across the whole of the Government machine (DETR, 1998c, p. 23). The combined effect of these new instruments of government has yet to be evaluated.

In England, government departments were combined to form the new Department of the Environment, Transport and the Regions (DETR), and it is this department that spearheads policy and action on sustainable development. It sponsors research, for example on sustainability indicators (DoE, 1996), and produces good-practice reports, for example on strategic environmental assessment (DoE, 1993) within the area of sustainable development; but it can be argued that this combined department provides just one minister in the Cabinet, thus making policy development and integration more fragmented than is the situation in Scotland, as will be discussed in a later section in this chapter.

Relevant to our 'equity' theme, the New Deal, including the Welfare to Work programme and the Environmental Task Force, was launched and a new Social Exclusion Unit was set up; its 'work will help all our people to share in the development of more sustainable communities' (DETR, 1998c, p. 3). Various consultation papers have been issued by the new government; here we note two that have particular relevance for our topic. *Our Healthier Nation* (Dept. of Health, 1998) recognises the relationship between health and the quality of the environment and the social and environmental causes of ill-health, offers a commitment to targeting inequalities in health, proposes a UK strategy to tackle poverty and social exclusion and has clear implications for urban policy and local transport policy. A parallel Green Paper was issued by the Department of Health in Scotland (Scottish Office, 1998a).

Spring 1998 saw the dissemination of a consultation paper on a revised UK strategy for sustainable development (DETR, 1998c). Identifying a number of themes, including 'how to promote sustainable communities for people to live and work in'(p. 5), the link between equity, environment and health is recognised. Following a period of consultation, a new UK Sustainable Development Strategy is expected about the end of 1998. An Integrated Transport White Paper appeared in summer 1998 (DETR, 1998d).

The Scottish context

This section presents a more detailed view of how one kingdom in the UK, Scotland, is pursuing the aims and delivery of sustainability. Scotland has some 10 per cent of the UK's population on 25 per cent of its land mass, with the majority living in the central belt. Otherwise, it is predominantly a rural country with an extensive coastline; it includes many islands. Scotland's environmental character, particularly in physical terms, has been summarised in a recent 'state of environment report' (SEPA, 1996). The socio-economic and environmental challenges facing the country have been succinctly highlighted by the AGSD (Advisory Group on Sustainable Development) 'vision statement' (AGSD, 1997, pp. 13–15).

In international terms we may note two particular features. Firstly, it is the UK,

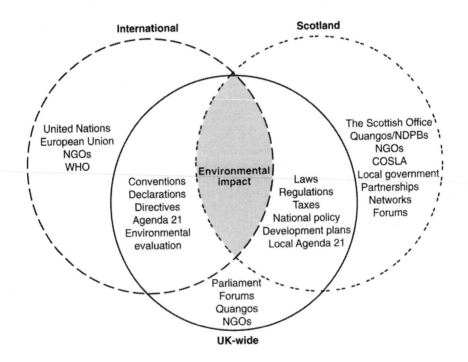

Figure 11.2 Spheres of influence upon environmental impact in Scotland pre-devolution.

not Scotland, which is a member-state of the EU. Scotland's influence in Europe is pursued through the representations of the UK government. Secondly, a number of NGOs and national partnerships achieve representation at international gatherings in their own right; for example, Friends of the Earth Scotland was present at the Rio+5 conference in 1997. Figure 11.2 demonstrates the interaction of the three spheres of influence – international, Scottish and UK-wide – and how they interact to produce an impact on the Scottish environment, defined here as including economic, environmental and social/community factors. On the formation of a Scottish Parliament in 2000, we will need to revise the diagram, since Scotland will then produce its own laws, as will be defined by the forthcoming Scottish Parliament Act. Thus the UK sphere of influence will diminish, as Scotland achieves more control over its own policy development. Box 11.3 offers more detail by listing the key Scottish bodies, operating at national level that are relevant to pursuing more sustainable development in 1998. Currently the Scottish Office has a key role. The minister with responsibility for sustainable development in Scotland states that 'the integrated nature of the Scottish Office enables a quite different approach to sustainable development in Scotland, simplifying simultaneous delivery of concern for the environment, economic growth and social progress' (Sewel, 1997b, p. 4). The Sustainable Development Unit was created in 1994 to co-ordinate future progress. Also in 1994, the Secretary of State established his own Advisory Group on Sustainable Development (AGSD) (Lindsay, 1996).

There are also a range of quangos, NGOs and voluntary groups concerned with pushing forward sustainability issues. 'Forward Scotland', launched in May 1996 as an independent company, is governed by a board of directors whose chairman is appointed by the Secretary of State. With 'towards a sustainable future' within its full title, it is 'committed to making sustainable development work' (chairman of Forward Scotland, in Scottish Office, 1998b, p. 29). Receiving its core funding from the Scottish Office, its purpose 'is to work in partnership with communities and organisations of all kinds in Scotland towards sustainable development' (Forward Scotland, 1997, p. 2). Four principles guide its operations: working to empower, enhancing integration, seeking partnership and 'playing fair' or supporting equity. For this organisation, sustainable development 'means seeking ways of working together to improve the quality of everyday life through practical actions while respecting the environment' (Forward Scotland, 1997, pp. 8–9). Figure 11.3 demonstrates the diversity of its partners and associated issues upon which practical action is offered.

Forward Scotland's 'model of sustainable development' as presented in Figure 11.1 demonstrates commitment to aspects relating to both equity and health – indeed several of its hundred or more projects relate to these issues. It initiated the Sustainable Communities Project which is one of our case studies later in this chapter.

At the local authority level, the Convention of Scottish Local Authorities (COSLA) represents the interests of the thirty-two unitary councils in Scotland. Through its committees and groups on environmental issues and sustainable

---

Box 11.3 **Key Scottish government bodies, quangos, NGOs and partnerships relevant to sustainable development, at 1998**

- Central government level includes:
  Secretary of State for Scotland
  The Scottish Office (five ministries and Sustainable Deveopment Team)
  Secretary of State's Advisory Group on Sustainable Development (offers independent advice to Secretary of State) (House of Commons, 1997b, 1997c)
  Proposed Scottish Parliament: the John Wheatley report (John Wheatley Centre, 1997) considers its role for sustainability

- Quangos/NDPBs (statutory bodies) include:
  Scottish Enterprise and local companies/development agencies
  Scottish Environmental Protection Agency (SEPA): see statutory guidance on sustainable development (Scottish Office, 1996) and *State of the Environment Report* (SEPA, 1996)
  Scottish Homes and housing associations
  Scottish Natural Heritage (SNH): National Heritage (Scotland) Act 1991 gives responsibility for achievement of sustainable development

- A private charitable company:
  Forward Scotland: (largely funded by the Scottish Office)

- Local government includes:
  COSLA (representing the thirty-two unitary authorities): committees/groups/publications on sustainable development, and see COSLA (1997)
  Local Agenda 21 Co-ordinators' Network (see Sewel, 1997b)

- Non-governmental organisations include:
  CBI in Scotland
  CVS Environment
  FoE
  Poverty Alliance
  Royal Society for the Protection of Birds
  Scottish Environmental Education Council
  World Wildlife Fund

- National partnerships/networks include:
  Rural Forum Scotland (long-standing)
  Scottish Anti-Poverty Network
  Scottish Contaminated Land Forum
  Scottish Environmental Forum (from 1992)
  Scottish National Rural Partnership
  Scottish National Transport Forum (from 1997)
  The Scottish Biodiversity Group (from 1996)

development, and frequently working in partnership with central government and the quangos, it issues best-practice advice on the Local Agenda 21 process (see, for example, COSLA, 1997). To complement COSLA's work, a Local Agenda 21 Co-ordinators' Network was established in the mid-1990s to support the co-ordinators employed by each of the thirty-two Unitary authorities.

Sustainability policy is linked with urban and rural regeneration policy at local government level, as expressed in 'New Life for Urban Scotland' (Scottish Office, 1988), which emphasises the importance of partnership. The Programme for Partnership publications (e.g. The Scottish Office, 1995a) explicitly recognised the links between social, economic and physical aspects within the review of the urban regeneration policy. The White Paper on policies for rural Scotland (Scottish Office, 1995b) emphasised sustainable development, which was developed further in a later discussion paper on the needs of rural communities in a development strategy for rural Scotland. Consideration has also been given to applying the principles in remoter areas (Wightman, 1996). Furthermore, the new ideas for national parks in Scotland 'will be built around the concept of sustainable development . . . paying attention to the need to balance conservation and the protection of the environment with economic and social development' (Sewel, 1997b, p. 2).

Recent consultation papers attempt an integrated approach, particularly in relation to health, equity, social exclusion, transport and sustainable development all within a 'best-value' context. See the key Scottish Office publications listed in

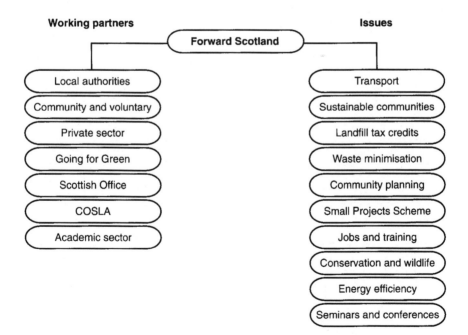

Figure 11.3 Forward Scotland: range of partnerships and issues.
Source: Derived from information supplied by Forward Scotland

the bibliography for an illustration of the range of initiatives. These policies are emerging and, as yet, have received limited evaluation of achievement. Every local authority has made a commitment to Agenda 21, and some have transformed that commitment into action and more elaborate programmes than others. The holistic approach of Local Agenda 21 offers a challenge to authorities, which are traditionally organised on sectoral lines. Several authorities are developing arrangements that cut across departmental boundaries. All have appointed a Local Agenda 21 co-ordinator; they are located in various departments, with chief executives, departments, environmental health and planning departments being the most common.

Key NGOs are listed in Box 11.3, along with other key partnership, community, and business organisations working for sustainability. Besides contributions to international forums, Friends of the Earth Scotland (FoE) published a substantive report (FoE Scotland, 1996). Developing the concept of 'environmental space' to provide a way of quantifying the changes sought, the report points to wise use of resources and the need to make dramatic reductions in the consumption of raw materials and the production of waste. Other publications cover various issues such as energy, poverty, atmospheric pollution (see, for example, FoE Scotland, 1995; Revie, 1997b; SAPN, 1996; SEF, 1997). Every authority in Scotland has made a commitment to Local Agenda 21: some examples are given below.

Partnership studies

Context

There are hundreds of Local Agenda 21 initiatives being pursued across Scotland. Here just four have been selected to illustrate contrasting examples of innovatory projects. Table 11.2 summarises the key characteristics of each partnership project before we review them in turn. Study 1 is a national pilot umbrella project with locally based initiatives, linked to housing policy; Study 2 describes an innovative sustainability process within the capital city; Study 3 is pre-Rio in origin, has a European dimension, and a focus in the sparsely populated areas of Scotland; and Study 4 relates to an energy project and the setting up of the first sustainable development company on the Scottish island of Islay.

Partnership study 1: sustainable communities programme, Scotland

The sustainable communities programme is aimed at enabling real communities to tackle sustainability at home, to find out what living as a sustainable community would be like. . . . I hope that we can extend this programme more widely as it is fundamental to identifying a sustainable future for the whole of Scotland.

(The Minister for Agriculture, the Environment and
Fisheries at the Scottish Office; in Sewel, 1997b, p. 6)

Table 11.2 Summary characteristics of partnership studies

Project	Characteristics
Sustainable communities project	Aim: to establish how communities of different socio-economic structures can gain sustainable improvements to their quality of life; national pilot project led by Forward Scotland in local partnerships; ten local projects, each with professional project officer, in diverse urban and rural communities; associated evaluative research programme.
The Lord Provost's Commission on Sustainable Development for the City of Edinburgh	Aim: to assess the challenges facing Edinburgh and provide recommendations that will help create sustainable development within the city and its interdependent region; within context of a city which has adopted an environmental strategy, and embarked on various other environmental initiatives; Independent Commission set by the city to report, within a year, to the Lord Provost; project is unique in UK, providing a new model of partnership; received evidence from hundreds of witnesses; report offers recommendations to each major institution and sector in the city.
CADISPA (conservation and development in sparsely populated areas)	Aim: to develop frameworks and models for sensitive rural development practice which focus on local people's interpretation of the concept of sustainability in rural development; commenced in Scotland in 1988, co-ordinated by WWF, and partly funded by EU; now a network of projects in peripheral rural regions in six European countries, set to expand its Scottish operations; linked to a university team, who have published various papers, including a model approach.
Islay Energy Project (and Sustainable Development Company)	Aim: to increase the island's energy efficiency, to contribute to regenerating the local economy and to integrate related projects; Islay is a peripheral Scottish island with a fragile environment and economy; importance of whisky distilling to the island's economy; leading to establishment of Scotland's first sustainable development company with the backing of public, private and voluntary bodies.

In Scotland, Forward Scotland was initiated as part of the UK government's response to the Rio Earth Summit and Agenda 21, and aims to get individuals involved in environmental action and discussion. As part of the Going for Green umbrella approach, 'there is a UK-wide sustainable communities programme which aims to establish if, and to what extent, communities of different socioeconomic structures can gain improvements to their quality of life' (Forward Scotland, 1997, p. 22). The network of sustainable communities in Scotland is led by Forward Scotland with support from Going for Green, FoE Scotland and the Scottish Office.

Launched in 1997 as a pilot project with an initial funding of 50,000, projects have been established in five authorities: Glasgow, Edinburgh, Stirling, Renfrew and South Ayrshire. Two communities are selected in each authority, normally contrasting areas, one of which is relatively affluent and the other more deprived. For example, in Stirling the two communities are Kippen, a relatively affluent settlement with little unemployment, and Fallin, an ex-mining settlement, where unemployment and associated deprivation are high. In Glasgow, however, both projects are in council house schemes of high deprivation, where there is a high incidence of unemployment, single parent families and dependence on income benefit. In Glasgow, the agreed objectives are to 'build community capacity and increase participation; attract resources and raise awareness about sustainable development' (Forward Scotland, 1997, p. 22).

For each project, Forward Scotland supplies the core team who work in partnership with the local communities and various agencies: 'each project has an officer responsible for its organisation, as well as good financial and political support from the relevant local authority' (Forward Scotland, 1997, p. 22). Aimed specifically at the contribution of individuals towards environmentally sound practice within the local socio-economic context, the officers work within communities to highlight problems and solutions. Practical solutions are sought to enhance the quality of living within available budgets. A research programme has been commissioned to evaluate the whole programme in Scotland; this will report after the completion of the pilot two-year programme. It could be that the Scottish Office, in partnership with local authorities, will wish to extend the programme into every authority in Scotland.

Friends of the Earth Scotland have also established a Sustainable Communities Project. This, the first, is based in Craigmillar in Edinburgh, and is adopting a different approach whereby, without a dedicated professional, the community is helped to work on issues of their own. The editor notes that this project is significant for its involvement of people from a variety of social backgrounds. In particular, it links working-class housing-area policy with environmentally sustainable initiatives – a combination that is often dismissed in England for a variety of classist and organisational reasons.

Partnership study 2: The Lord Provost's Commission on Sustainable Development for the City of Edinburgh

The vision behind all the Commission's recommendations 'is that by the year 2020, Edinburgh should be a sustainable, human, prosperous, innovative, efficiently managed, and compact city' (press release for the Lord Provost's Commission, 13 July 1998). Edinburgh is Scotland's capital city where the Scottish Parliament will open in 2000. It is a medium-sized city, with a population of just under 500,000 people; relatively compact and densely populated city, with a strong architectural heritage and a generally good natural environment. Despite economic growth during the 1990s, a 'major problem facing the city is the socio-economic divide between rich and poor', particularly the deprivation experienced in many peripheral estates (McLaughlan, 1997, p. 2).

Echoing the work of the former Lothian Regional and Edinburgh District Councils, the City of Edinburgh Council has been establishing policies and projects relating to sustainable development for several years. It has produced an environmental strategy (City of Edinburgh Council, 1998); addressed sustainability issues, including health, through its Urban Regeneration Programme (Capital City Partnership, 1996a, 1996b); and promoted radical transport initiatives, including reclaiming pedestrian space on the Royal Mile and on Princes Street, the implementation of intensive bus priority 'Greenways' and introducing city-centre car clubs. It is also the first Scottish city to have introduced car-free residential areas (City of Edinburgh Council, 1996; Hazel, 1997; Lothian Regional Council, 1996; DETR, 1998d). Its initiatives have been recorded in research programmes and practical guidelines on urban sustainability (see, for example, Jenks *et al.*, 1996; Mazza and Rydin, 1997; Pedestrians Policy Group, 1996). Thus Edinburgh has been pursuing integrated, innovative and experimental programmes on the path towards developing its Local Agenda 21 programme and the achievement of sustainability.

It was within this context that the Lord Provost's Commission on Sustainable Development was established in 1997 (see Figure 11.4). It was set up by the Edinburgh Environmental Strategy Core Groups (now the Edinburgh Environmental Partnership), a body which brings together key players within the city to promote an environmental agenda. Set up like a Parliamentary Commission to hear evidence, with a budget of £30,000, it was supported by the City Council and Lothian and Edinburgh Enterprise Ltd (LEEL). As the first project of its kind in the UK, its remit was 'to take evidence from all sectors of the community, to review the available information, to identify where additional information is needed . . . and to advise the Lord Provost on such changes in policy as might be necessary to enable the city of Edinburgh to achieve a socially, economically and environmentally sustainable pattern of life' (Lord Provost's Commission, 1998, pp. 4–5).

Fourteen commissioners from various walks of life were appointed to this independent body. All commissioners 'volunteered' their time and the group included just one councillor, the chair of Edinburgh's Local Agenda 21 sub-committee. An external assessor was also appointed. Officially launched in January 1997, it started work that April. The Commissioners identified six issues as priority areas for action:

- poverty and social exclusion;
- energy and global climate change;
- waste management and resource use;
- quality of the neighbourhood;
- transport and air pollution;
- land-use management and planning.

Figure 11.4 The Lord Provost of Edinburgh, Eric Milligan, meets with members of his Commission on Sustainable Development for the City of Edinburgh.
Source: Photograph courtesy of the City of Edinburgh Council.

Witnesses representing the NGOs, education, community groups, business and industry, local government, national government and quangos, were invited to offer written evidence. Some seventy were subsequently invited to be questioned by the Commissioners at a series of seven public hearings: this formed the core of the evidence-gathering process. In addition, several hundred community organisations were asked their views and the Commissioners participated in a number of open public events throughout the year (Lord Provost's Commission 1998, p. 5). Their 88-page report was published in July 1998. It reviews the evidence presented, including the perceived barriers to the achievement of a more sustainable city, and offers 127 recommendations for action by local government, government, universities, business, industry and community organisations. The report concludes with three overall recommendations for establishing:

1   a Lord Provost's Charter and Award; the Charter, to which organisations could sign up, to facilitate progress towards desired goals; the Award, as part of the Charter, to be presented to an organisation that had made most progress towards sustainable development in the previous year;
2   a Sustainable Edinburgh Partnership; the proposal is to convene an annual Community Partnership on Sustainable Development linked to the Edinburgh Festival;

3   an education and awareness-raising campaign; the lack of understanding of sustainability emerged as a major issue in evidence presented to the Commission, thus it recommends the development of a unified campaign.

(The Lord Provost's Commission 1998, pp. 83–4)

Edinburgh looks well on the way to become a model for other UK and European cities in their development of policy and action towards sustainable development. As part of that programme, and through engagement with the major players in the city, the Lord Provost's Commission demonstrated a willingness genuinely to allow people to submit evidence as it sought to reach a conclusion on the best way forward in promoting economic regeneration in a sustainable fashion. This case study provides an example of an innovative approach to the Local Agenda 21 process. The editor adds that this is an example in which both conventional town planning issues and mattters of equity and environmental sustainability are integrated. Readers may also find it instructive to compare this case study of applying Agenda 21 principles to Edinburgh with the case study of Glasgow in chapter 10, in which WHO principles were applied. Both integrate 'social' issues alongside 'physical factors' within a broader context, respectively sustainability and health.

Partnership study 3: CADISPA

The CADISPA project 'offers local people substantial technical and practical help to enable them to achieve their local agenda' (Forward Scotland, 1997, p. 13). CADISPA (Conservation and Development in Sparsely Populated Areas), describes itself as 'The agency for sustainable development in the rural community'. Pre-dating Rio, when the programmes started CADISPA was unique in its approach of considering environmental and economic concerns, conservation and development, as parts of one equation, not conflicting issues. Box 11.4 outlines the origins and work of CADISPA at a European level.

The first CADISPA projects in Scotland were on the Isle of Uist, in the Outer Hebrides, and in the Flow Country of Caithness (both starting in 1989). As pilot projects, they aimed to 'find out how high environmental concerns ranked in the lives of locals, develop innovative education programmes that respond to the area's concerns by using local examples, and utilise case study materials that would justify extending the project to other European countries' (CADISPA, 1995b).

By 1997 CADISPA was actively involved with ten community development groups in six areas of sparse population throughout Scotland: the islands of Coll, Colonsay and Tiree, Easdale Island, Gatehouse of Fleet and the Burghhead Amenities project. The total sum of rural sustainable development currently being put in place now stands at 2.75 million (Fagan, 1997a, p. 22). Local people are central to the development process, and CADISPA, acting as a facilitator to empower local communities, considers it helps 'local people interpret and implement Local Agenda 21 from within their own local development agenda' (Fagan, 1997a, p. 2).

Box 11.4   The origins and work of CADISPA

In 1988, WWF-UK wanted to start a project that would provide entire rural communities with an education that would allow them to make decisions about the use of the environment with insight and understanding. WWF contacted the University of Strathclyde. The project assumed the acronym CADISPA and its seed was planted in windswept Scottish highlands and islands.

CADISPA was extended to Italy and Spain in 1990, and Greece and Portugal in 1992, establishing a European network that is co-ordinated by WWF International. Still expanding, the programme will soon support projects in Sweden and Albania. As a non-profit making umbrella group for several organisations co-ordinated by WWF and partly funded by the European Union, CADISPA supports dedicated local project executants to develop activities that address the specific conservation and development problems in their areas.

CADISPA sites have usually remained rural and underdeveloped for a reason: poor soils rule out large-scale farming, inaccessibility and lack of infrastructure prevent tourism and other forms of development, absence of exploitable minerals keeps mining interests away. Small changes can have a great impact on these often fragile environments. The CADISPA challenge is to help rural people come up with ways for them to use local resources to grow economically, and protect the environment and preserve their cultural identity at the same time.

Although each CADISPA region's culture, language and economy require different approaches, all the areas were selected according to common characteristics:

- Sparsely populated areas are subject to constant emigration, further population decrease, and an ageing population.
- All areas are rich in biodiversity, but have an already damaged, fragile and precarious environmental balance that is threatened by damaging plans such as dams, roads, bridges, intensive farming, forestry, hunting and mass tourism.
- Local residents are poor economically, but rich culturally. They resent outside agents of decision, due to historical circumstances, and generally have low capacity for self-defence and self-determination.
- Local environmental organisations or WWF offices are active at a grassroots level and ready to collaborate.

Under the CADISPA umbrella there is a constant exchange of information and experience between the country teams involved. CADISPA's long-term goal is to produce a model that empowers local communities to support economically viable development initiatives that also conserve the natural environment and the cultural heritage. The CADISPA team thinks education is the key: working with teachers and students, training and motivating local residents and policy makers will lead the way to sustainable development.

(CADISPA, 1995a)

Concerned with sustainable communities and the interpretation of Local Agenda 21 through community development, CADISPA defines sustainability to include:

- economic regeneration of a kind which is administered by local people for themselves;
- an environmental sensitivity which is addressed in all the development decisions local people may take;
- social and cultural sustainability promoted through the building of secure organisational structure and the celebration of cultural uniqueness and diversity.

(Fagan, 1997a, p. 3)

This puts 'capacity building', the developing of skills and understanding which enable people to take action, at the heart of the CADISPA definition of sustainability and, in part, accounts for why all projects last a minimum of three years.

Current sponsors include foundations and trusts, the Argyll and Island Local Economic Development Company and Forward Scotland. Examples of projects include a new community hall in Coll and Colonsay, a Gaelic art and language archive in Tiree and the Gatehouse Youth Drop-In Centre. Considerable expansion of activity is planned, from ten to thirty-five projects, stretching across Scotland. The CADISPA approach to community development is described in a research publication (Fagan, 1997b). The editor adds that this is a case study concerned with rural, sparsely populated areas, but, as in many rural areas (as discussed in chapter 4) 'social planning' issues related to gender, age, class and deprivation are just as prevalent (albeit on a smaller scale) as in big urban locations.

Partnership study 4: the Islay Energy Project and Islay Sustainable Development Company

The belief is the Islay Energy Project could become a dynamic, real demonstration of how energy – and renewable energy in particular – could be developed to provide sustainable benefits for an entire community.

(Rutherford, 1998, p. 27)

Lying off the west coast of Scotland, Islay is part of the Inner Hebrides. The island is mostly rather low and rolling, being a mixture of green farmland and peat moss. The population is just under 4,000, with about half living, in equal proportion, in the two main settlements of Port Ellen and Bowmore. The Gaelic language is still widely spoken, but mostly among the older generation. Islay's economic base rests upon the island's natural resources and reflects traditional and diverse indigenous industries: agriculture, fishing and manufacturing, which is dominated by whisky distilling, the latter being a significant source of income to the British exchequer. However, the largest number of islanders work in service industries, with tourism making a substantial contribution to the island's economy. Young people have been tending to leave the island for opportunities elsewhere.

Islay is linked directly to the mainland, a journey of some thirty miles, by a daily roll-on, roll-off ferry service; there is also a daily air service. As with all islands, the physical separation and isolation generate factors relevant to land-use planning which operate differently, or not at all, in a mainland situation; many aspects of land use are vitally interrelated to each other in their demands on land and in their impact on the prosperity of the island population; there is a special dependency on sea transport, which imposes constraints and costs that pervade almost every aspect of island life; furthermore, outside influences, for example economic changes and political decisions at all levels, can have a massive influence at island level and form the chief limitation on local land-use planning. The development of land cannot be dissociated from its ownership: on Islay it is largely held in extensive private estates and there has been a significant influx of second-home owners. A fuller description of the nature of and challenges facing Scottish rural economies, including Islay, may be found in a number of publications (see, for example, Argyll and Bute DC, 1985; Baxter and Usher, 1994; Selman and Barker, 1989; Scottish Office, 1995b; SNH, 1996).

The island depends upon imported energy in the form of electricity, coal or fuel oil for most of its energy needs. A comprehensive study into energy use and resources on the island concluded that 'there was significant potential for alternative renewable energy generation and that the whole community should aim to increase its energy efficiency' (Rutherford, 1998, p. 26). It was from these findings that the Islay Energy Project was launched; the study partners represent a wide range of agencies and community groups, including six companies involved with whisky production. Seen as a step in wider moves to regenerate the local economy, it aims to integrate related projects on the island and has three key objectives:

- demand-side management: to ensure that existing resources – oil, electricity, coal and peat – are used in the best possible way;
- renewable energy: to explore other sources of energy, such as wind or wave power;
- environmental balance: to maintain the right balance between economic development and the protection of a precious natural environment and strong local culture.

(Rutherford, 1998, p. 27)

Progress is being achieved. A site has been identified for wind generation, although it is controversial, owing to environmental objections on the grounds that the proposed wind cluster would constitute a danger to geese. A small-scale experimental plant has demonstrated that wave-power generation is feasible, and proposals for a commercial demonstration device are under development. The creation of new jobs on the island will help to protect the economy and make it possible for young people to stay.

The Director of Scotch Whisky Production, United Distillers and Vintners, has acknowledged that, 'The Islay Energy Project has been an important catalyst in moves towards the sustainable development vision of the future' (Rutherford, 1998, p. 28). The issues being focused on by the partners in this island community relate to renewable and sustainable energy, transport, employment, local amenities, land use, natural heritage and green tourism, and bear a close relationship to the government's rural policy development. Whilst conflicts of interest on the island are not new (Selman and Barker, 1989), the industries, the islanders and the environmentalists are recognising that visions need co-ordinated support: the first sustainable development company in Scotland was launched in March 1998. It has the backing of various partners, including the Scottish Office, Argyll and Bute Council, the local population, local trusts, Forward Scotland and the whisky industry. It is anticipated that ' the formation of this company should not only lead to sustainable developments in local life, but could become a model that would be of interest to the rest of Scotland and beyond' (Rutherford, 1998, p. 28). This example is of interest in that it shows a strong economic/employment component to the application of sustainability principles in a small community.

Case study résumé

To conclude the review of these four contrasting partnership approaches, we can note that each demonstrates:

- the importance of partnership approaches, often with many and diverse partners;
- an inclusive approach with the community, involving concerted attempts, frequently as facilitators, with diverse communities, so that they have the opportunity to shape agendas and control further action, i.e. capacity-building to engage the community and keep the momentum going;
- innovative approaches which may provide models for elsewhere, thus informing policy delivery;
- a need for funding from various sources, nearly always including substantive voluntary effort;
- a part of the Local Agenda 21 process for the area in which they are located;
- an indirect relationship to the national and local planning process.

While much work remains to be done, it may be concluded that, whilst not without criticism, these partnership projects celebrate the triple top of addressing the three concerns of sustainability (defined in both socio-economic and environmental terms) of equity and, to a lesser extent, of health.

Conclusion: towards the 'triple top'

Policy on social progress is one of the three arms of Scottish policy. It includes promoting health-enhancing and environmentally sustainable policies, along with social equity. Significantly, the term 'social planning' itself is seldom used, but we have seen that the concepts, particularly relating to equity and deprivation issues, are being pursued in partnership within the urban regeneration and rural development programmes. Increasing attempts are being made to integrate these three arms through sustainable development projects within the Local Agenda 21 process. As yet, many of these are pilot projects and experimental, but increasing activity suggests that Scotland may be poised to make substantive progress in policy development to achieve a more sustainable society.

However, significant barriers to the achievement of sustainable development have been identified. These are noted below with examples of their implications:

- Lack of knowledge and awareness:
  the whole population of Scotland needs to become 'sustainable-development literate';
- Financial constraints and funding:
  recent emphasis on 'short termism' whereby increasing numbers of people are working on short-term contracts, making them cheaper to employ, but increasing insecurity
  charges for natural resources are not always related to consumption
  identified need for cash incentives, alongside concerns about competitiveness;
- Infrastructure and co-ordination:
  structure of funding for economic regeneration
  lack of integrated public transport system
  lack of ready mechanisms to deliver the environmental, economic and social aims of sustainability together;
- Regulation and legislation:
  a key driving force for environmental concern in industry
  the structure of existing regulations rarely offers indicators on whether an organisation is acting sustainably
  UK energy pricing is socially regressive;
- Leadership:
  not enough leaders in civic and business life have embraced sustainability and given it priority;
- Departmentalism:
  even with a corporate approach, lack of integration can continue;
- Lack of available data:
  both to inform the community of directional movement towards sustainability and for more technical measurement, e.g. on resource consumption and waste disposal;
- Cultural/institutional issues:
  wastes are seen as something to be disposed of, rather than as a potential source of income (see, for example, Lord Provost's Commission, 1998, p. 17; FoE, n.d., p. 24).

Understanding impediments to progress should assist in deducing ways of overcoming such obstacles. Our studies of partnership demonstrate the significance of the planning function at national level, and the importance of partnership to achieve social progress at the local. Just as 'environment' proved to be a problem word for policy-makers, so is the word 'sustainability' challenging: there is no doubt that the concept includes equity, health and planning – we may yet be able to celebrate the triple top in the twenty-first century.

Part IV

Cultural perspectives on planning

12 Cultural planning and time planning*

Franco Bianchini and Clara H. Greed

The relationship between culture and urban planning

Franco Bianchini

I want to reflect on the relationship between culture, cultural production and the cultural sector more generally, and urban planning, starting from the need to widen the agenda and move towards a more holistic notion of planning - a need which is increasingly recognised by planning theorists and practitioners.

Every period in the history of the development of cities seems to need its own forms of creativity. Urban planners this century have been especially influenced by the creativity of engineers and scientists, who in Victorian Britain had responded to the problems of overcrowding, mobility and public health generated by the Industrial Revolution by building terraced housing, railways and sewage systems. Such focus on 'hard infrastructures' was taken up and developed by planners in the twentieth century, for instance with the construction of inner ring roads to ease access to central city areas for motorists and, more recently, with the installation of CCTV systems as part of attempts to reduce crime in town centres. Today there is an increasing awareness that such physical and 'scientific' approaches can only be part of the solution to urban problems. Urban crime, for instance, cannot be solved simply by physical control without reinforcing pride in the locality and mutual responsibilities, both in city centres and in residential neighbourhoods. It is also clear that cities will not become more ecologically sustainable if we do not address how people mix and connect, their motivations, and whether they 'own' where they live and change their lifestyles appropriately. In short, what urban planners also need today is perhaps the creativity of artists, more specifically of artists

*The first part of this chapter is based on a speech given by Franco Bianchini on 'Rethinking the relationship between culture and urban planning' at 'The Art of Regeneration', a conference held in Nottingham, 1996 (Matarasso and Halls, 1996), with additional material from *The Creative City* (Landry and Bianchini, 1995). It is an example of an alternative way of looking at urban issues from the perspective of cultural planning research, which transcends the social/spatial dichotomy that has dogged the progress of British town planning. Readers may also be familiar with Franco's work on the related issue of 'time planning', which is discussed in the second part of this chapter by Clara H. Greed, drawing on a range of European material following discussion with Franco.

working in social contexts, like the British 'community artists' and the French cultural animateurs. This is the creativity of being able to synthesise; to see the connections between the natural, social, cultural, political and economic environments; to gauge impacts across different spheres of life, and to grasp the importance not only of 'hard' but also of 'soft' infrastructures. The latter are the social and cultural networks and dynamics of a place, which include the daily routines of working and playing, local rituals, traditions, ambiences and atmospheres, as well as people's sense of belonging and of 'ownership' of particular localities, buildings, institutions and activities. A knowledge of soft infrastructures and of how to use them is crucial for planners for successful policy implementation.

There are other limitations to the way planners operate. Some are very well known, such as a relative lack of sensitivity to the culture of cities interpreted as history and heritage, which was especially evident in the 1950s and 60s. One good example of the critique of planners by urban historians is provided by Mark Girouard who, in *The English Town*, writes:

> I went back to Huddersfield and found that its heroic streets of mills had long since vanished under a ring road . . . I saw how ruthlessly two-thirds of the centre of Worcester and most of the centre of Gloucester had been mangled; walked from the station to the Shire Hall through the corpse of what had once been Chelmsford; discovered how Taunton had destroyed in a year or two the town centre so carefully and creatively formed in the 18th-century; wandered in Liverpool past gutted buildings and over the acres of desert once covered by 18th century squares and terraces; and wept in the screaming desolation of central Birmingham.
>
> (Girouard, 1990, p. 7)

During the 1950s, 60s and early 70s much of what remained of walkable town and city centres with historic buildings, interesting vistas and overlapping uses was dismembered and distorted by new inner ring roads, multi-storey car parks and subways, which are unfriendly and dangerous, especially for women.

The erosion of urbanity through insensitive planning is continuing now, although clearly not in such extreme forms. Many of the indoor shopping malls built in British town and city centres in the last decade perpetuate the mistakes of the 1960s. They are perhaps fine on one side, which is integrated into one of the main shopping streets, but on the other three sides they are impenetrable fortresses. Their blank walls blight the surrounding environment and create a new edge, a new barrier, further shrinking the town centre.

There are also serious problems in the training of urban planners. It has been largely forgotten that urban planning and design is a form of art. It is intrinsically linked with other forms of cultural production. It is about the production of a city, and the city is in part an artefact. There was more of a continuum between architecture, urban planning, urban design and art in ancient Greece or in Renaissance Italy–societies which in this respect are still looked at as possible models.

The existing training of planners – in disciplines including regional science, urban geography and economic policy – needs to be enriched, through the input of disciplines in the humanities like art history and philosophy, perhaps by blending a northern European approach with a southern European style which places greater emphasis on the aesthetic and historical appreciation of cities. Unless such new forms of training are developed there will continue to be problems, because planning is conceived from a narrow conceptual and epistemological basis from which it branches into very complex issues such as local economic development, tourism promotion, and place marketing.

A more intimate relationship with culture and cultural production is not the solution in itself. We need, however, a critique of urban planning from a cultural perspective, in order to pour it into the policy-making pot to begin to produce something new. To be able to do this it is important to be clear about what notion of 'culture' we will adopt for this redefinition of urban planning. As Raymond Williams (1961) argues in *The Long Revolution*, there are three broad categories in the definition of 'culture':

1   the aesthetic notion of culture as 'art', 'the body of intellectual and imaginative work in which human thought and experience is variously recorded' (R. Williams, 1961, p. 41);
2   'culture' as the cultivation of the spirit and the mind, 'a state or process of human perfection, in terms of certain absolute or universal values' (ibid.) – a definition that is obviously linked with the concepts of 'learning' and 'education';
3   the anthropological notion of culture as 'a particular way of life, which expresses certain meanings and values not only in art and learning, but also in institutions and ordinary behaviour' (ibid.).

The anthropological definition of 'culture' is the most useful for the purposes of developing a critique of urban planning from a cultural perspective. The idea of linking culture intended as 'a way of life' with planning has been developing in the US and Australia (see, for instance, Mercer, 1996) and in the last decade has begun to be considered in some urban areas in Britain. The local authority in Kirklees, for example, produced a document in 1991 whose primary objective was to respond to a multiple crisis, the roots of which they identified in 'the erosion of local culture'. According to the local policy-makers, the dominance of multiple shops and the growth of out-of-town shopping centres was beginning to erode the cultural distinctiveness of the different towns that make up Kirklees, including Huddersfield, Dewsbury and Batley. The dominance of the car was furthering the dispersal of urban functions. The crisis of staple, traditional industries was also contributing to eroding local culture, as was the privatisation and domesticisation of leisure, with people spending more time watching videos and TV and engaging in other forms of home-based free-time activities like DIY. Finally, the crisis of local government under a massive centralisation of power and resources by central government since 1979 was another factor in the crisis of local culture and identity.

Kirklees' response was innovative (Wood, 1996). The local authority adopted a 'cultural planning' approach, intended not as 'the planning of culture' – a dangerous and probably impossible undertaking – but as a type of planning that was sensitive to local cultural specificities. 'Cultural planning' differs from traditional arts, media, heritage and recreation policies for several reasons.

The first is that traditional cultural policies tend to be based upon a narrow definition of 'culture', which by and large is not 'culture' as 'a particular way of life', but as 'art'. Second, they tend to have a sectoral focus, i.e. they concentrate on the development of separate cultural sectors or forms. There are thus specific policies for drama, dance, literature, the visual arts or the crafts. In cultural planning, on the other hand, there is a broad definition of 'culture' and a territorial focus. The key is to see how the broad spectrum of cultural resources can contribute to the development of a place, whether neighbourhood, town or region.

To explore further this idea of cultural planning, you have to imagine a diagram in which the centre is occupied by the pool of cultural resources available in a certain place (see Figure 12.1). These resources include:

- arts and media activities and institutions;
- youth, ethnic minority and occupational cultures;
- the cultures of different communities of interest;
- the heritage, including archaeology, gastronomy, local dialects and local rituals;
- local and external perceptions of a place, as expressed in jokes, songs, literature, myths, tourist guides, media coverage and conventional wisdom;
- the natural and built environment, including public and open spaces;
- the diversity and quality of retailing, leisure, culture, eating, drinking and entertainment facilities and activities;
- the repertoire of local products and skills in crafts, manufacturing and services.

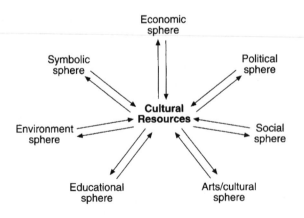

Figure 12.1 Towards a 'cultural planning' approach?

If you place this set of resources at the centre of the table of policy-making, you can imagine two-way relationships between these resources and any type of public policy – for example, economic, transport, environmental, physical planning, social or educational policy. There are two relationships. One arrow, so to speak, goes from each sphere of policy-making to cultural resources, meaning that we have to take seriously those resources in developing social, educational, economic or any other type of policy. There would then be another arrow going from the cultural resources to each sphere of policy-making, implying that policy-makers have to learn something from the processes of thinking used by people working in the field of cultural production – i.e. the production of meanings, images, narratives and ideas. The main characteristics of these processes can be summarised in six groups of attributes:

1   holistic, flexible, lateral, networked;
2   innovative, original and experimental;
3   critical, enquiring, problematising, challenging the status quo and questioning;
4   people-centred and humanistic;
5   cultured, and critically aware of history, of the culture of the past;
6   open-ended and non-instrumental.

If we accept this way of conceptualising the relationship between culture and urban planning, we need to rethink the roles of the cultural manager, of the arts and cultural development officer and of the cultural policy-maker in organisations like local authorities. These professionals would become – in addition to knowing how to develop a sector of cultural production – the gatekeepers, and the mediators between the sphere of cultural production and different areas of public policy-making.

The potential of this reconceptualisation becomes perhaps clearer by looking at a specific policy like place marketing. By and large, place marketing policy-makers come from a background in advertising or marketing. They have had experience of marketing commercial products like brands of drinks, cars or clothes and have then tried to transfer these processes to a city. The political pressure to achieve results can mean that they may have just six months to develop a new marketing strategy for a place as complex as a city with 400,000 inhabitants.

It is impossible to market Liverpool, Bologna, Lyon or anywhere else in the same way as you market yoghurt. To return to one of the the six attributes of the thinking processes used in cultural production listed above, city marketing would need to become more holistic, interdisciplinary and lateral, by being based on consultation and research and by involving people who know about the city at a deep level; it would include teamwork between planners, architects, historians, anthropologists, psychologists and sociologists. Otherwise there is the danger of ending up with a stereotypical strategy, which can be basically the same in Nottingham, Rennes or Milan. On the contrary, place marketing strategies should be shaped by the detail of the culture of the place.

I also think, at the risk of sounding a little elitist, that there is an issue in terms of the cultural skills of urban policy-makers. We have reached the limits of the instrumental approach to culture-based urban regeneration developed in the mid- to late 1980s, when local cultural life was used as a tool for 'selling' cities to tourists and investors, but was rarely understood to the point where planners and other policy-makers would change their own ways of thinking and operating. A critique of the cultural skills of politicians and policy-makers has been aptly made by Justin O'Connor, who characterises the prestige cultural flagship projects of the 1980s and early 1990s as 'a product of buying in the most staid, unadventurous and downright bad design' (O'Connor, 1993, pp. 17–18). O'Connor adds that 'this says something about the power of money' – for example, the power of the construction industry striking deals with politicians – 'but it also says something about the real lack of cultural skills amongst the local elites in politics, business and "development". This is of course a widespread problem in British business and management, where they stand as cultural illiterates next to many of our European competitors' (ibid., p. 18). For example, in Manchester, says O'Connor:

> the skills and the imagination were there, but they were passed over as the regeneration coalition of the '80s was drawn towards the big development capital whose only vision of prestige was Beethoven's Fifth and the mock luxury of Laura Ashley and plastic stretched awnings. We would just need to compare Manchester's Olympic bid to that of Barcelona. Bob Scott (leader of the Manchester Olympic bid) tells us this is a 'classic bid', by which we should read safe, unadventurous and riddled with unimaginative design.
>
> (ibid.)

Will many of the prestige buildings erected in Britain as part of the urban regeneration strategies of the 1980s and '90s stand the test of time? What about the public art that has been sited over the last two decades in key public spaces in many cities in Britain? Is it of good quality? Does it reflect the culture of the place? Is it properly integrated into urban design and planning strategies? Or is it something added when the development has been completed, sometimes as a desperate attempt to apply, as the saying goes, lipstick on the gorilla?

I want to conclude by looking at some of the things that need to happen if urban planning is to be redefined by taking seriously into account the attributes of the process of cultural production listed earlier.

I have already said that we need new forms of training. There are serious limits to the existing training of physical planners, of city marketing officers and of the cultural managers themselves. A narrow training in arts administration is inadequate for cultural planners, who also need to know about political economy and urban sociology, about how cities work (as societies, economies, polities, and eco-systems, as well as cultural milieux) and of course about physical planning itself, otherwise they cannot influence it.

Equally, there is a need for new forms of research, not focusing exclusively on the economic impact of cultural resources and cultural policies as they did in the late 1980s and early 90s but also on their impact in terms of social cohesion, organisational capacity (Matarasso, 1997), and the development of the innovation potential of cities.

New structures for policy-making are required. These should be much more integrated and overarching, bringing together departments and working on issues and themes, rather than on the basis of policy empires and departments. For example, if the issue is that the town centre is dead at night, in order to develop a response we need to bring together policy-makers responsible for lighting, policing, transport, cultural animation and physical planning.

It is also important to develop a new culture in local authorities and other public policy-making bodies as well – a culture more oriented towards innovation and not afraid of failure, or at least able to distinguish between competent and incompetent mistakes, and accept the former as something that may contain the seeds of future success. The use of ideas competitions – currently largely restricted to architecture and urban design projects – could be extended to projects dealing with crime, with urban transport, and with the design of cultural policies themselves.

More generally, a useful critique of mainstream approaches to planning and policy-making is provided by the former Irish Minister of Culture, Michael D. Higgins – who, in addition to being a politician, is also a poet and a philosopher. He said in his speech at the conference 'The Economy of the Arts' in Dublin in 1994:

> For too long . . . financial institutions have used their hegemony to set limits to policy in other areas, constantly diminishing the cultural space in which so much radical or innovative thinking is possible. One result has been a dire impoverishment of social philosophy. We no longer seem to be living in countries but in economies . . . *Homo economicus* feels justified by his products, whereas play is concerned with means rather than ends, with the quality of an action rather than its result. Hence the major contradiction of our economic arrangements: that a society based on the negation of the play-element presents itself as uniquely able to deliver play – but only as an experience of consumption.
>
> (Higgins, 1994, p. 2)

Higgins argues that we must recover an ancient wisdom which teaches us that play, 'far from being a deviation from the workaday norm, is the basis of all culture' (ibid.). According to Higgins, it is important to restore a humanist dimension to policy-making:

> minds which for decades have ceased to ask why they do what they do have doomed themselves to mere systems maintenance . . . the sheer volume of facts to be digested by the students of so many professional disciplines leaves little time for a deeper interrogation of their moral worth. The result . . . has been a generation of technicians rather than visionaries, each one taking a career rather than an idea seriously.
>
> (Higgins, 1994, p. 3)

In short, Higgins develops a powerful argument for a reintegration of play into planning and policy-making, which would help us overcome narrowly economistic thinking, and move towards much richer, more sophisticated strategies as the basis for more (economically, socially, environmentally and culturally) sustainable cities.

Time and the city

Clara H. Greed

> Il y a un temps pour tout et chaque chose sous le ciel a son heure
> (Ecclesiastes ch. 3, verse 1; quoted in ARU, 1997 p. 95)

'What time is this place?' revisited

The British planning system has been obsessed with utilising spatial, land-use control as a means of ordering cities and dealing with urban problems. This has often been at the expense of wider cultural considerations, as discussed above by Franco Bianchini. In this section written by the editor, based upon discussion with Franco and an expanded version of an article published in *Planning*, (Greed, 1997a), 'time planning', another form of planning, which derived from Italy and is, arguably, more sensitive to cultural issues, is discussed.

Many the problems which confront modern cities are not primarily spatial in nature, but are generated by the way we organise and divide up time for different uses and activities within cities – that is, they are temporal. Little conscious attention appears to be given to this issue, we simply accept the existing organisation of urban time as 'natural' or 'normal', according to the mores of our culture. If time is organised at all it is done in a fragmented and restrictive manner by individual organisations such as businesses, factories, schools and public service providers, with an input by magistrates and licensing authorities in the case of entertainment premises. Because the present system seems so 'natural' inevitably people become alarmed by government proposals to 'change time', for example to abolish British Summer Time (daylight saving) or to introduce the European Work Time Directive (which, although intended to shorten working time, has been a cause of genuine concern, as it could be used to permit a twelve-day working week!)

Lack of temporal organisation results in functional inefficiency. For example, during the morning rush hour, roads are unable to accommodate present-day volumes of traffic, resulting in congestion, pollution and, nowadays, increasingly, 'road rage', as well as time simply wasted sitting in traffic or standing at bus stops. In contrast, at other times of the day many roads are used below capacity, road space is wasted, and public transport is intermittent. The effect in social terms is that many people, especially parents with caring responsibilities, find it inconvenient, difficult or downright impossible to get their children to school and themselves to work in the morning. The problems are both temporal and spatial. People who work full-time and who do not have the benefit of that anachronism 'the non-working housewife' also find it difficult to find a time in which to do their shopping and to undertake simple tasks like going to the Post Office or the other 'little things' that are so undervalued in our society but are essential to existence: getting to the shoe-repairer, dry cleaner, vet, dentist and library.

Many of these problems could be ameliorated by the introduction of time planning, following the example of several European cities. This should not be confused with the currently trendy, and arguably male, concept of the '24-hour city'. This concept seems to many cynics to be a legitimation of the '24-hour booze', based on a glamorised, 'Brits abroad' holiday image of the Continental city as consisting entirely of bars, nightclubs and cafés. Images of Ibiza and all-night clubs such as Hedonism spring to mind. A bit of 'urban design' and 'culture' may be thrown in to justify this parody of urban life (cf. Montgomery, 1994, who, in subsequent correspondence – in response to Greed, 1997a – argues this was never his intention). Such concepts contribute little to solving the wider structural and practical problems of the city in respect of residential, employment and transportation policy. They have nothing to offer those workers with 'carer' responsibilities who have no time to stay on in the town centre drinking the night away, as they have to get back home and start on the 'second-shift' of cooking, housework and childcare.

Whilst the introduction of 'all-night' clubs, even '24-hour clubs' such as the famous Lakota, in Upper York Street, Bristol might satisfy the needs of those fancy-free young people who love to go clubbing (including some students if they can afford it), does it really contribute to the well-being of the city as a whole? Demands for extensions to closing hours of clubs, bars and pubs might be welcomed by the police and magistrates, who have found that this can 'dilute' the lager lout factor, and spread the problem of everyone spilling out at once onto the streets of the city at closing time. But local residents who actually live in the surrounding areas may consider that music throbbing away all night, plus all the inevitable disturbances associated with large numbers of people using any public entertainment facility, are factors at odds with seeking to maintain a semblance of family life within the inner city. In contrast, time planning – as will be illustrated with reference to examples from Europe, especially Italy – is far broader, is concerned with the needs of the whole urban population, and has implications for every aspect of the way in which cities and their land uses are planned. Time planning is discussed from a full range of perspectives in a special issue of *Les Annales de la Recherche Urbaine* on 'Emplois du temps' (*ARU*, 1997).

Italian time planning

Italy pioneered time planning, originally in its northern cities. This section of the chapter is based, in part, upon papers presented at a recent Eurofem conference centred on time planning, hosted by the Valle d'Aosta municipality (the GLC of the Alps!) in association with planning colleagues from Turin University such as Carmen Belloni, a key time-planning specialist (Belloni, 1996), with delegates from throughout Europe. Local government reforms introduced in 1990 (under Article 36, 142/1990) gave elected city mayors the powers to formulate 'time plans' in association with business and school representatives, and to 'change hours', that is, to rationalise the efficency and convenience of the times of opening of shops, schools, places of work and public services: surely a form of social town planning! Time planning has now become widespread across over eighty municipalities, including model examples such as Modena in the region of Emilia Romagna. Other cities which have implemented measures include Pistoia in the region of Tuscany and Catania in Sicily, which pioneered experimental time planning in 1988. Naples and surrounding villages have also experimented with time planning, as have Milan and Rome to some extent, in urban situations where spatial planning is not as strongly developed as it is in Britain.

These time reforms have been the result of pressure from feminist politicians, sociologists and planners, many of whom have links with the Democratic Party of the Left (Bianchini, 1995). Their programme and achievements in 'temporal planning' may be compared to the 'spatial planning' breakthroughs made by the women's committee of the GLC in London and other progressive metropolitan authorities in the 1980s. In Italy the increase in the numbers of women with children going out to work, currently around 60 per cent, combined with the peculiarly Italian system of opening hours for offices, shops, factories, governmental offices and banks, involving siestas and intermittent opening, makes it increasingly difficult for both families and individuals to function, because nothing is ever open when people want it. In Italy there is very little part-time employment as we know it, making it even more difficult for women to fit in with the demands of school hours. There is also a heavy emphasis on traditional patriarchal family values, which disadvantages working women. In addition, there are major traffic problems in many Italian cities, so quite apart from the social dimension it has been seen to be in the interest of everyone, regardless of their gender, to plan time as well as space, if only to increase road capacity by spreading traffic volume more evenly across the whole day.

Women in Italy have found it impossible to get their children to school by 8.00 a.m. or earlier and themselves to work at the same time, especially with the growth of cities increasing journey times between land uses. (It is still chiefly women who are assumed to be responsible for children.) In the afternoon schools finish at 1.00 p.m. and there is also the problem of the siesta, with shops closing until around 4.00 p.m. In addition, some offices, schools and factories still have six-day weeks and others have peculiar shift patterns, with afternoon working spreading into the evenings in some cases. The system breaks down once there are no traditional 'mamas' left in the kitchen, or mothers or grandmothers at home all day supposedly 'with nothing else to do', to provide the support and make all this work.

In Italy at present there is also a move to reform the administrative, bureaucratic system of public sector offices, which is a problem for any member of the public trying to get permits, official forms, passports or identity cards without having to take a day off work to do so. Multinational companies moving into northern Italy have been trying to create work structures more similar to those found in North America. Italian companies such as Fiat in Turin have also been experimenting with flexible shiftwork, in association with the trade unions. To date, this has resulted in over 100 permutations, which in turn, interestingly, have reduced the demand for and viability of public transport usage in the city, thus adding to private car use and congestion. Clearly, time planning is a complex field requiring sophisticated solutions: it cannot be done on a partial basis. As the 1960s exponents of a systems view of planning taught, everything affects everything else (albeit neither 'time' nor 'gender' were ever mentioned in those theories), and cities must be planned holistically. Integral to time planning is better streamlining and provision of services for childcare and the elderly, plus detailed improvements to make health clinics and hospitals more accessible, as have been undertaken in Modena. Of course, all this has to be tied in with transport planning and public transport providers' total involvement. Time planning extends the municipal planners' powers, but also requires increased community involvement in, and commitment to, the planning process.

Employees in many Italian organisations nowadays are given a choice of a full free day, rather than just Sunday or Saturday. Many women prefer a free weekday as the shops, dentists, doctors and public services are open and they get time for themselves, whereas many men prefer Sunday because of the football (Juventus in Turin) and because they don't have to fit in the other responsibilities of housework, homemaking and the millions of little errands and duties that women still have to undertake. It is better for the consumer if offices are open on Saturdays or even Sundays. It should be remembered that Saturday and Sunday have always been working days for women, particularly in households where cooking, childcare and housework are never-ending tasks. British delegates at Aosta thought that local government in Britain had much to learn from this, but the question of 'weekend rights', and 'free Sundays' is a thorny issue in Unison trade union circles. Other concepts included 'time banks', which are part of a bartering system whereby people offer a service, such as doing shopping for the elderly, in return for childcare. The 'time bank' idea is growing and there is talk of running offices and schools on this principle too, breaking down home/work divisions and job demarcations. All this ties in with wider trends for more flexible 'human' work structures and the breakdown of gender-roles.

German time planning

There have also been 'time planning trials' in northern European countries – where, arguably, somewhat more rigid cultural attitudes towards time prevail – for example, in Germany, as discussed by Dietrich Henckel (Henckel, 1996). Interestingly, the reasons he gives for German attempts at time planning are the problems of traffic congestion, changing employment patterns, the need for efficiency and post-unification reconstruction, and associated 'stress', rather than primarily a concern with the social planning issues of gender and the related 'double shift' of work and home duties for women. However, women in Germany have pressed for many years for reform of shop opening hours, which are some of the most restricted in Europe. Women planners in Germany have also found that the Green movement has been particularly supportive, as have sustainability initiatives in general, of both gender issues and time planning (Spitzner and Zauke, 1996; Spitzner, 1998). The legal scope of EU Environmental Impact Analysis is increasingly being interpreted in Germany and Scandinavia to include a consideration of social factors as part of the assessment critieria, to an extent not found to be the case in Britain. Clearly, time planning is seen by environmentalists as a highly sustainable strategy, as time is the ultimate non-renewable resource – albeit invisible and priceless. The introduction of flexible working patterns, often designed more for the convenience of the employer than the employee, means that a full-time, 'job for life' is becoming more scarce in an increasingly competitive, globalised economic situation. National governments need to increase the scope of their planning powers to maintain an acceptable way of life for their citizens, even though many pundits declare the nation-state is powerless, its government is becoming redundant against such global economic forces.

It has been found in reunified Germany that only 30 per cent of workers nowadays work a traditional 9–5 (or 8–4) day. Yet urban transportation and land-use planning systems are still geared up to meet the needs of the now mythical full-time, car-borne, and presumably male worker: and there are still just as many cars travelling during the morning rush hour. But are they all commuting to work? It has been found in Britain (for example, by Mayer Hillman) that up to 30 per cent of morning rush hour traffic nowadays is generated by people taking children to school. The reasons for this growth include matters of distance, safety (road and personal), wider school choice and the need to drive to fit in the journey to school with the journey to work in the very limited time available to working parents. Interestingly, some Scandinavian countries have experimented with a range of 'walking to school' alternatives, all of which demand both staggered time planning in terms of start times for work relative to school, and dedicated staff supervision.

Over the day as a whole, only around 35 per cent of all journeys are work related, and many are composite journeys (Barton, 1998b, p. 135, Table 8.2). Indeed, the car becomes essential for many people for whom cities of disparate land-uses and unsynchronised working hours have made it necessary to move as quickly as possible between two places that they actually need to be in simultaneously – but the result is that everyone grinds to a halt. There is a need, therefore, for both time and space planning, and for a change in our conceptualisation of the scope and objectives of town planning, so as to accommodate all of people's needs as human beings, not just as 'workers' and 'commuters' but as 'carers', 'homemakers', 'shoppers', 'parents' – and through the whole of their lives as 'children', 'grown-ups' and 'elderly people', too. In particular, women want planning for practical realities and necessities, rather than for abstract economic goals. The Eurofem group has advocated the concept of 'planning for everyday life'; it wants a city of short journeys, less land-use zoning and mixed-use districts, with more emphasis on the local distribution of shops, local facilities and amenities, and with better childcare provision as advocated in Scandinavia, for example (as set out in Skjerve, 1993), and promoted in Finland under the leadership of Mrs Siirpa Pietikäinen, Minister of the Environment.

New times in Britain?

In conclusion, it is argued that many of the principles of time planning might be applied in Britain, too. There has certainly been academic, if not implementatory interest in the use of time in cities in the past, as evidenced by popularity of the work of Pred and of Hagerstrand from Sweden, both of whom, incidentally, foresaw the changes the 'information revolution' and new technologies would have on working practices (Hagerstrand, 1968; Pred, 1977). There has been research in Dublin, for example, on the relationship between gender and time in shaping urban space (see Bhride, 1987, on time-budgeting). A temporal master plan would go towards solving many of the problems of childcare, 'latchkey kids', stress, congestion, and impossible schedules in our cities. Harmonising all opening hours and travel timetables would certainly benefit people in Britain too. A holistic, inclusive approach would go far beyond simply a 'flexi-time' or staggered rush hour approach, both of which only address time problems associated with employment structures. Indeed, the Office of National Statistics is to undertake a time-use survey for the first time, as part of a Eurostat EU wide time study, based on a national sample of participants filling in a 24-hour diary. Women in London more than twenty years ago advocated co-ordination of all office, factory, shop, public building, transportation and school hours: perhaps now 'the time is right'.

13 Planning for tourism in towns and cities

Rob Davidson and Robert Maitland

Tourism planning

This chapter discusses the impact of large numbers of people from outside the area on our towns and cities, and ways of managing tourist numbers – surely a form of social town planning in all but name. In the closing decades of the twentieth century, with tourist and leisure activity expanding globally, more and more people have been confronted with the reality that the towns and cities where they grew up, and in which they now live and work, have also become magnets for visitors – from neighbouring towns or from the other side of the planet, or both.

Cities can become recognised tourist destinations in a variety of ways. For many, it is a deliberate decision, often driven by necessity, as traditional industries decline and new economic activities are needed to create employment and prosperity. Locally elected representatives may choose to market and develop the city for tourism, based on its natural, built and cultural resources. Equally, tourism can play a key role in regeneration projects within cities, promoting vitality and playing a key role in mixed use developments (Maitland, 1997).

Other cities (and city districts) are 'discovered' when entrepreneurs or developers from today's dynamic tourism industry, with its voracious appetite for expansion, see in them an opportunity for development. There then may follow the construction of hotels, conference centres, shops, restaurants, recreational facilities and tourist attractions, whose appeal may only partly depend on the attractiveness of the place in which they are located. Many of these facilities are used by local people and day visitors as well tourists spending a night in the destination.

Finally, in Western, developed countries in particular, the general population's determination to exercise its freedom to travel to wherever it wishes results in many other towns and cities becoming tourist destinations – in some cases, if not against their will then at least without their active encouragement. A population ever more mobile and ever more restless, well informed, sophisticated and curious is increasingly given to seeking out new experiences in places previously untouched by tourism. Consequently, places offering the right type of resources are increasingly finding that they are attracting substantial numbers of visitors (Davidson and Maitland, 1997).

The expanding global appetite for travel and tourism is, of course, not only resulting in the creation of new destinations. Tourism has long been an established industry in many 'traditional' destinations such as seaside resorts, attractive country areas, historic towns and cities, where it is often a major economic activity. Tourism's continuing growth has, therefore, also fuelled increases in the number of visitors to many such long-established destinations, in particular those that have some unique quality or resource that retains, through time, its appeal for the travelling public. These are frequently public resources, used by but not owned by the tourism industry. They include natural resources (such as landscape), built environment resources (such as historic buildings) and socio-cultural resources (such as festivals or cuisine).

Whether we are considering places where tourism is relatively new or places that tourists have been visiting for hundreds of years, it is clear that visitors have the potential to bring substantial benefits to the community and its residents. It is equally clear that, left uncontrolled, tourism can wreak damage to the very fabric of the destination, through the many well-documented negative impacts that it can bring about through the over-exploitation of public resources, and the damage that can be done to the economy or to the general quality of life of residents.

As tourism has spread to more destinations, and as more places have seen tourism as a crucial part of how they make their living, it has become clearer that to be considered successful it must not only provide a satisfactory experience for visitors, but also maintain or improve the quality of life of residents and protect the local natural, built and cultural environment. Equally, the tourism industry itself, in particular the private sector, must be rewarded, in the form of healthy profits, for the risks it takes in investing in commercial facilities at the destination.

It is now widely accepted that this requires effective planning. A recent UK report jointly authored by the National Planning Forum (NPF) comprising the tourism industry, local authorities and other environmental bodies, argued that:

> Tourism development needs to be planned for and managed in order to ensure that the positive benefits are not outweighed by the negative impacts. It needs a robust planning framework and effective management mechanisms.
>
> (NPF, 1998, p. 8)

This consensus around a broad role for planning is comparatively recent. Until the 1990s, most planning for tourism was primarily oriented towards controlling development to protect the environment, or the intensive marketing and promotion of destinations and the satisfaction of the needs of the tourists themselves. Typically, politicians and tourism planners had as their priority the maximisation of the financial benefits of tourism for their country, region or city, while the tourism industry did its utmost to attract as many customers as possible, in order to receive the greatest returns on its investments.

Archer and Cooper (1994) observe, however, that 'Fortunately, the climate of thought is changing, albeit slowly', (p. 88), with a greater awareness developing of the consequences of excessive and badly planned tourism expansion and the importance of the host population. Gunn (1994) also notes the move away from the strong preoccupation with *promotion* as the sole key to development, and the growing tendency for tourism planners to recognise several goals and interests, in addition to that of an improved economy: increased visitor satisfaction, integration with existing local social and economic life, and protection and better utilisation of destination resources.

Several factors have contributed to changing the general approach to tourism planning. Significant among them is the rise of environmentalism and sustainability as major global issues. As Goodall and Stabler (1997, pp. 279, 280) point out:

> evolution of the concept of sustainable tourism has paralleled that of the wider concept of sustainable development. . . . The currently dominant tourism-centric paradigm of sustainable tourism [means]:
>
> - meeting the needs and wants of the host community . . . in both the short and long term
> - satisfying the demands of a growing number of tourists and of the tourism industry . . .
> - safeguarding the . . . resource base of tourism . . .
> - maintaining or enhancing the competitiveness and viability of the tourism industry.

These concerns have stimulated interest in tourism planning and management as a way of ensuring that principles of sustainable development are observed as destinations develop, and as a way of attempting to reconcile the differing interests of the various stakeholders involved.

The trend towards sustainability in tourism is international and growing (OECD, 1996). One general consequence has been the trend away from the rigid 'grand design' master plan in favour of more flexible, integrated approach, with planning for tourism taking its place alongside other forms of planning in the locality. This has been particularly marked at the local destination level. Gunn (1994, p.17) notes the growth in interest in planning for tourism at the local destination level and the proliferation of destination plans:

> Now there is greater sensitivity to visitor capacity management, integration of tourism with the local society and economy, and protection of natural and cultural resources,as foundations for more and better travel experiences. New organisational mechanisms that foster collaboration on planning are appearing. More areas are seeing the value of planning that utilises input from local citizens as well as planning specialists and consultants.

The last point is very significant. Professionals involved in tourism planning having been increasingly moving from an elitist approach to greater involvement of all parties involved in and impacted by their decisions, reflecting the approach long established in UK land-use planning. Simmons (1994) suggests that this concern with participation has come about for two reasons: first, the recognition that the impacts of tourism are felt most keenly by those living in the local destination area and second, the realisation that community residents are an essential ingredient in the 'hospitality atmosphere' of a destination.

Any involvement of a wide cross-section of participants in the planning system will necessarily involve not only conflicting goals but incompatible perceptions about the tourism industry. One of the functions of tourism policies, then, is to attempt to reconcile the divergent interests of the different actors. Significant emphasis on partnership is therefore one of the distinguishing characteristics of tourism policies at this level. Van den Berg (1994, p. 174) makes this observation in his study of urban tourism in Europe:

> Tourist policy should reflect that complexity by taking into account the relations between the relevant elements of the private and public sectors . . . Partnerships, public–public, as well as public–private, are eminently suitable to meet in practice the need for overall quality and integration

This in turn gives professionals involved in planning an additional role to play: that of catalytic agents for new solutions to planning problems, resolving conflicts by bringing opposing factions together to agree on a plan which is acceptable to all sides.

This brief survey shows an emerging approach to tourism planning characterised by:

- concerns with principles of sustainable development. Although there is predictable disagreement about what this means and what (if anything) should be done, it has led to an emphasis on destinations and their development as the focus for planning (Goodall and Stabler, 1997);
- the potential of tourism in regeneration;
- concerns with the host population and their involvement in planning and management;
- an emphasis on the centrality of partnership arrangements in tourism planning and management.

Tourism planning, in the widest sense, is composed of all those plans, policies and strategies that have as their objective, or as one of their objectives, the guiding and shaping of the development of tourism at destinations. Some of these plans are statutory and enforceable by law; most are non-statutory, taking the form of agreements and guidelines for action. In the next section, we examine the complex process which constitutes tourism planning for towns and cities in the UK.

Planning for tourism in towns and cities

As Heeley (1981) has pointed out, tourism is not a self-contained policy area. It takes place at different geographical scales, and 'it overlaps with policy fields such as transport, conservation, rural development, and so forth, so that only a small proportion of the sum total of plans affecting tourism are exclusively devoted to it' (ibid. p. 61). For every plan dealing solely with tourism there are many in which tourism features as one element among several. For example, at the national level in the UK, tourism is an element in plans for the conservation and management of the country's forests, waterways, national parks, museums and many national monuments, while at the local level, tourism features, for example, in many individual counties', districts' and cities' economic development plans.

Tourism is one of the forms of development dealt with by the UK land-use planning system. At national level it is covered by PPG 21 (Tourism), but PPG 17 (Sport and Recreation) is also relevant, and there are strong arguments for developing integrated guidance on tourism, leisure, sport and hospitality. PPG 13 (Transport) affects tourism, and in 1997 the government made it clear that the PPG 6 (Town Centres and Retail Development) sequential approach should be applied to leisure as well as retail development (Select Committee on the Environment, 1997). (The sequential approach requires that suitable sites be sought in, successively, town centres and edge-of-town locations before out-of-town sites are considered.) Whilst 'tourism, leisure and recreation form one of the strategic topics which should be considered in Structure Plans and Part 1 of Unitary Development Plans, 'it has not in the past featured strongly in development plans . . . [and] the commercial sector in particular is concerned that the . . . emphasis is too negative' (NPF, 1998, p. 13).

The new emphasis on sustainability and the reinforcement of a plan-led approach may change this, but to date formal, statutory land-use planning processes have been only one element in planning and managing tourism. The rest of this chapter is concerned with some of the other measures used for the planning of tourism in towns and cities: entrepreneurial planning, local authority tourism strategies, local partnership arrangements and Visitor Management Plans.

Tourism and entrepreneurial planning

Land-use planning controls and guides development, and by doing so can act positively to make the most of a destination's potential and protect and enhance its resources. However, such planning is heavily reliant on private sector action for its effects, and without pressure for development, its scope is limited. In many areas of the UK for much of the 1980s and 1990s, there has been little or no development pressure: consequently planning has sought to *create* products and projects to achieve regeneration. Tourism has played a central role.

Whilst the 1980s saw a preoccupation with market forces and the 'enterprise culture' (Deakins and Edwards, 1993), growing awareness of the limitations of a purely market-led (and especially property-led) approach to development, together with concerns about the environment, sustainability and the resultant conflicts with development imperatives, saw a return to an apparently more plan-led approach with the introduction of the Planning and Compensation Act, 1991 (Thornley, 1991).

A consequence was the emergence of a more holistic approach to planning and regeneration, described by Stoker and Young (1993) as 'entrepreneurial planning'. It means understanding the requirements of the private sector so as to create a market and lever in private investment in particular areas. These may be within generally prosperous cities (for example Bristol Docks, where the Arnolfini and Watershed arts complexes played a crucial role), within cities coping with major economic change (Little Germany in Bradford) or in smaller towns. One way of characterising this approach to renewal is as developing a business-plan approach to area development – an approach that tries to combine long term vision from a community point of view with the requirements of the private sector. In this way, a greater concern with longer-range planning has begun to return, though in a different form.

Entrepreneurial planning works within the framework of development plans, but relies for its effect on an ability to influence private sector investment through marketing, promotion and subsidy. It requires more than investment in property and the physical environment. It sees area revival coming from a mixture of investment, with measures to address local needs and to recreate vitality and life in an area. This means that tourism and leisure, with their potential to attract people at all times of the day and to generate fun and festival, can have a central role to play.

Stoker and Young (1993) describe entrepreneurial planning in terms of 'flagship projects and schemes on big vacant sites'. They describe a five-stage approach to development, identifying five implementation tasks: sites, strategy, subsidies, spotlights and skills. Tourism and leisure activities can be crucial to this process. (Although Stoker and Young are not writing from a tourism perspective, six of the nine specific developments they quote to illustrate the entrepreneurial planning process are tourism or leisure schemes.) Tourism can be involved at the most crucial stages of the process. It can assist *land renewal* by providing potential uses for derelict historic buildings; museums or galleries are common *public investment projects* that can help to build confidence. As a growth industry with links to other activities, tourism or leisure uses can be a vital component in development packages designed to *initiate public investment*. The high visibility of such projects and their potential to attract widespread interest makes them important in *strategies* to promote developer confidence and an obvious target for *spotlights* – positive publicity. Since many projects will involve the arts or sport, they may attract specific *subsidies* which can be added to regeneration funding. Finally, some the *skills* required for entrepreneurial planning – such as marketing – may be well developed amongst those working in tourism and leisure.

Entrepreneurial planning, then, seeks to achieve regeneration through a process that combines public objectives and private sector requirements: inherent in this approach is the idea of partnership between public sector agencies, the private sector and the public. Elements of this approach have been more widely adopted. The NPF report referred to earlier (NPF, 1998) calls for tourism planning which has proactive strategies and produces a vision for the area – key ingredients of the entrepreneurial planning approach.

Local authority tourism strategies

Local authority (LA) tourism strategies provide a framework to guide local tourism development. They vary considerably in their scope, content and level of detail. Some are based on detailed research and analysis, highlighting specific action and initiatives. Others are simply general statements of council policy and intentions towards tourism.

Many useful insights into the formulation and content of LAs' strategies for tourism are provided by J. Long's (1994) study based on a survey of tourism officers. It showed that planning departments rarely have a leading role: 'although the work of many hands, these British strategies have more commonly been produced by Leisure Services Departments than by Planning Departments' (p. 17).

Three principal reasons were given for producing the strategies:

1   to draw different parts of the authority together in common purpose (especially important in those authorities that do not have a tourism department) – or sometimes to provide the basis for restructuring;
2   to provide a manifesto for tourism (especially in those authorities in which the role of tourism was not established) which could assist the search for funds;
3   to help to ensure effective co-ordinated action in pursuit of agreed goals.

Long favourably compares the strategies included in his survey with the strategic plans reviewed in a similar survey undertaken in Australia (Dredge and Moore, 1992), which were 'rooted in the traditions of land-use planning', whereas 'the British strategies . . . were more wide-reaching, notably also encompassing organisation and administration, promotion, information and development, Indeed, at their best, they seem to be rather closer to what Dredge and Moore have in mind as being desirable' (J. Long, 1994, p. 20).

The various strategies reviewed by Long were distinguished by the extent to which they were characterised by one of the following themes:

*   Development of the 'product':
    through improving infrastructure (roads, signs, etc.);
    through enhancing the attractions of the area;
    through changing/improving the accommodation base.

*   Promotion of the destination:
    management of tourism to protect the environment or interests of local residents;
    organisation/administration of tourism in the local area.

Another distinguishing factor identified by Long was his respondents' definition of good practice in their authority's tourism efforts:

The nature of good practice varied in part as a result of the different roles and relationships respondents saw for the public and private sector. A simple distinction can be made between those who saw their role as one that would establish the conditions that would allow the private sector to operate profitably, and those who were concerned to encourage the private sector to get involved in meeting desirable goals in keeping with the resort or destination. Yet others saw the role of the LA as that of honest broker, balancing the competing interests.

(J. Long, 1994, p. 21)

The prominence of partnership ventures clearly emerges from Long's research as a vitally important theme. Whether for the purposes of development or for promotion, partnership arrangements with adjoining authorities, individual companies and other agencies have enabled LAs to achieve much more than they could have done single-handedly. The next section describes two of the most prominent partnership arrangements entered into at the local level.

TDAPs and LATIs

In the 1980s, tourism planning at the destination level in the UK came to be characterised by the growing recognition that successful tourism plans require the participation of a variety of actors and interests for their successful execution. There was also a growing impatience with detailed and comprehensive tourism studies that are long on analysis but short on practical action. The trend has been therefore for shorter and more action-orientated programmes, often addressed to or involving a wider cross-section of interests. Such local tourism partnerships between the public and the private sectors have been promoted by the UK government and the national tourist boards, and have been widely established throughout the UK.

Turner (1992) regards the proliferation of partnership arrangements partly as a result of the prevailing politico-economic situation, with central government directing LAs to act as 'enablers', not 'providers' of services, and partly as a result of restrictions on local government spending. From the perspective of the private sector, he provides another reason:

On the private sector side, anticipated difficulties in achieving the desired rate of return from tourism developments, particularly from innovative schemes, has led private enterprises to look to local authorities either to contribute to capital costs or to offer favourable rental terms (where they are the landowners) in recognition of the wider economic benefits of new developments. There has, additionally, been a realisation that there needs to be a pooling of resources in what is a very fragmented industry in order to compete with spending on the promotion of overseas holidays. There is also a recognition, within both public and private sectors, that for many of the complex infrastructure problems which confront tourist areas, a simple solution is insufficient, or such a solution requires a partnership between the institutions which make up the local community.

(Turner, 1992, p. A85)

Organisational arrangements for local tourism partnership initiatives sponsored by the English Tourist Board (ETB) during the 1980s were known as TDAPs – Tourism Development Action Programmes. They were replaced in 1990 by their successor LATIs – Local Area Tourism Initiatives. Between 1984 and 1990, twenty different areas within England were designated as TDAPs by the ETB, in consultation with local interests. P. Long (1994, p. 481) lists the following characteristics shared by these initiatives:

- They seek collaboration and co-operation between a wide range of public and private sector organisations located in, or having an interest in, their areas' tourism . . .
- They are action-oriented, with the programme emphasis on implementing initiatives rather than on prolonged and detailed research and strategy formulation.
- They are comprehensive and integrated in approach, encompassing the range of interrelated aspects which affect tourism in the programme area. Consequently, they normally include development, marketing, information and environmental advisory and training initiatives.
- They are corporate in approach and involve objectives and work programmes which are shared both among and within organisations. This is particularly important when tourism-related activities are the functions of several different local authority departments and where the programme area transcends administrative boundaries.
- They are short-term programmes with a limited duration, usually three years, with the aim of establishing sufficient momentum for progress to be sustained in the longer term, based on local resources.

The distinguishing feature of LATIs is their narrow focus on urban districts, such as the London boroughs of Greenwich and Islington, Leeds Waterfront and Manchester's Castlefield district. The partners in all LATIs are:

- the local authority, or several local authorities within the LATI area;
- local private sector operators and associations such as Chambers of Commerce and travel associations;
- the relevant regional tourist boards;
- the relevant national tourist board.

Other organisations directly involved vary between each LATI, but have included:

- National Park authorities;

- urban development corporations and economic development and enterprise agencies (themselves partnership organisations);
- local community groups and associations, e.g. civic societies and parish councils.

Formal organisational arrangements for TDAP/LATI administration and management have generally involved a steering group comprising representatives from partner organisations and a working group (or groups) responsible for programme implementation. Such programmes have few direct employees, but rather involve staff on secondments from partnership organisations or personnel allocated to partnership responsibilities as part of their jobs with partner organisations. 'Nevertheless [LATIs] can be considered as organisations in their own right, with their own identity, management and objectives, albeit with a high degree of dependency on the sponsoring partners and with a possibly limited life span'. (P. Long, 1994, p. 484).

Visitor Management Plans and town-centre management

Visitor Management Plans (VMPs) are non-statutory plans focusing on particular impact issues, such as the problems caused by tourism pressures on infrastructure. They may be seen as a type of tourism strategy developed in response to the growing concern about the environmental and social impacts of tourism and the wish to maximise the benefits of tourism for the destination. As such they have a vital part to play in achieving sustainability for tourist destinations.

Visitor management is a comparatively new discipline. Human (1994, p. 221) distinguishes VMPs from earlier plans and strategies:

> During the 1980s, tourism strategies and development plans were viewed almost as a panacea for unemployment by LAs who heeded advice to take advantage of the economic benefits of the industry. In the more environmentally-aware 1990s, VMPs are the order of the day.

In 1990, the UK Government Tourism and Environment Task Force, co-ordinated by the ETB and involving a wide range of industry and local authority representatives, developed a definition of visitor management as 'an ongoing process to reconcile the potentially competing needs of the visitor, the place and the host community'. The aim of visitor management is to achieve a harmony between these three elements, taking into account the following considerations:

- the interrelationship between the physical environment, residents and visitors;
- the need to balance the requirements/demands of each group;
- external factors (which are often beyond the control of the destination) such as national/EU legislation, or physical or financial constraints.

Visitor management is sometimes regarded simply as finding a solution to vehicular and pedestrian congestion. But the true meaning of the term is much more than this. It is an approach that destinations can use in order to 'integrate visitor activity within the long term planning and day to day management of a town/city and to maximise the benefit of tourism, including its contribution to town viability and vitality' (McNamara, 1996, p. 17, citing M. Grant, 1996). In preparing VMPs, tourism is considered *in relationship to the functioning of the destination as a whole.*

The term 'visitor management' itself is ambiguous, as it is not the activities of the visitors themselves that are being managed, but the destination's services and facilities. Visitors are not under the direct control of the destination's managers (local authorities, for the main part), but many services and facilities are, and these can be managed to indirectly 'control' visitors' activities through suggestion and persuasion. The English Historic Towns Forum report *Getting it Right* (EHTF, 1994) lists the possible benefits that successful visitor management can bring to the destination:

- enhancing the visitor experience;
- enhancing the town's reputation as a destination to visit;
- creating a quality environment in which to live and work;
- maximising the economic opportunities of tourism;
- minimising the associated impact, particularly on local people;
- encouraging more profitable overnight stays;
- encouraging off-season visits to extend the length of the season and to reduce peak season pressure;
- reducing wear and tear at sensitive sites;
- identifying the necessary capital and maintenance costs for town centre management;
- strengthening a local sense of civic pride.

Elsewhere M. Grant (1996) adds three more benefits of visitor management: contributing to the vitality of the town (especially in the evenings); supporting existing facilities and encouraging new facilities that will benefit residents and visitors alike (e.g. retail, sports, arts and leisure facilities); and encouraging and creating a demand for the conservation of specific heritage facets, as well as contributing to the costs of this conservation.

To translate these benefits and objectives into action, a typical VMP encompasses the destination's marketing plan and various aspects of the development of the product, including interpretation and signage, transport policy and traffic, visitor services (toilets, guides, etc.), attractions (including shopping) and conservation and environmental improvements. It also aims to identify residents' needs and to gain their commitment to the plan. Implementation calls for action from a range of partners, and achieving consensus around common aims is a key element of the process. These partners typically include a wide range of local authority departments directly or indirectly concerned with tourism, private sector representatives (e.g. Chambers of Commerce, traders' associations) and conservation groups (e.g. civic trusts), as well as residents' representatives.

Within urban destinations, many aspects of visitor management are closely related to wider town centre management (TCM) issues. TCM has been developed in response to the range of problems facing traditional town centres as they have increasingly come under pressure from several directions. Grant *et al.* (1995) argue that many have suffered from competition from out-of-town developments, decline in the numbers of central area residents and threats to the evening and night-time economy. Some, on the contrary, are victims of their own success, suffering from the pressures of excess demand, with consequent pedestrian and traffic congestion. Tourism is often a major contributor to this pressure. The consequences of these problems, if they are allowed to develop without any form of strategic intervention to tackle them, are a poorer environment, reduced services and a fall in viability and vitality.

TCM is 'a comprehensive programme by public authorities, private sector interests and voluntary organisations, which aims to improve the standards of facilities, environment, convenience and safety in town centres' (Grant *et al.*, 1995, p. A22). Its goals, according to the Association of Town Centre Management, are to:

- achieve a competitive edge;
- improve the management of the public realm;
- satisfy the aspirations of all town centre users.

(ibid.)

The emphasis is on creating a *partnership*, a *shared vision*, and *direct action* to achieve improvements. Typically, schemes develop strategies around the environment, access, security, promotion and retailing vitality. Practical projects include business development and skills training, paving schemes and control of fly-posting, closed-circuit TV, tackling aggressive begging, access for people with disabilities, and advance information on car parking.

The main benefits of effective TCM are described by Grant *et al.* (1995, p. A23) as the following:

- It emphasises the uniqueness of the place, builds consensus, partnership and civic pride, and encourages return visits.
- It attracts development, encourages private sector investment and support, and encourages joint operator initiatives, such as quality staff training.

There is a clear overlap between tourism concerns and those of TCM, since:

- Visitors and local people use many of the same facilities, attractions and services (e.g. car parks, museums or restaurants), which are often concentrated in town centres.
- The tourism industry itself reaps distinct advantages from TCM, since it integrates tourism with other town centre functions as a mainstream feature; it improves the general image of the town, giving a practical marketing advantage; it is cost effective – the concerns of TCM and visitor management

overlap, and a great deal can be achieved without duplicating research, staff, liaison and investment.
- At a policy level, both call for building shared aims, consensus and co-ordinated action – across LA departments; between different parts of the tourism industry; broadly, between public and private sectors; and between local people, visitors and commercial operators.

Finally, there is a definite complementarity between plans for TCM and the other types of plan we have examined in this chapter:

> There is a vital role for VMPs, tourism and economic development strategies and Local (land-use and transport) Plans to complement and work with TCM to ensure that the full benefits of tourism are achieved with a minimum negative impact on the immediate environment and local communities. TCM will often provide the enabling mechanism to meet, in part, the objectives of these other structures by implementing practical projects.
>
> (Grant *et al.*, 1995, p. A24)

Conclusion

Planning for tourism involves a multitude of actors and the implementation of a wide variety of measures. However, no single plan, policy or strategy of the types examined in this chapter can provide destinations with either a panacea for tourism-related problems or a get-rich-quick key to instant prosperity through tourism. Because of the great variety of destinations concerned, the very wide range of measures necessary and the vast number of different areas of responsibility that come into play, no single scheme can tackle all the issues. There is no off-the-peg package which will guarantee all destinations success through tourism. It is, however, possible to identify a number of features that have characterised the emerging approach to planning for tourism:

- the importance of vision, around which a consensus for action can be built;
- emphasis on partnerships: public–public, public–private and private–private;
- strategies that are wide-ranging and cross-departmental;
- strategies that are action-oriented, non-statutory, and have short time horizons – three years is common;
- emphasis on partnership and on project-based organisations, often with few or no permanent staff, for delivery.

In many respects, this parallels developments in other fields of urban management, notably regeneration, especially where there is an emphasis upon community and user involvement. Arguably, 'tourism planning' is another valid version of 'social town planning', albeit from a more entrepreneurial perspective than some other aspects.

14 Culture, community and communication in the planning process

Jean Hillier

Planning for cultural diversity

The activity of local planning policy-making is often underlaid by a complex interplay of actors, diverse values and mind-sets. The ways in which different planners and community groups perceive themselves, their own, or others' geographical places and other groups respectively, reveal cultural differences of self- and place-identity, of discourses and values. The decision-making processes involved in land-use planning are temporary points of fixation in time at which actors bring together their different representations. The question of which identity and representation is dominant will be the result of social negotiation and conflict, and the effectiveness of communication. Actors from different cultural lifeworlds, using different discourses, may talk past or at each other rather than with each other. As an approach to building a shared understanding I refer to the Habermasian concept of communicative action.

A participatory, communicative approach to decision-making is never easy: the greater the number of actors involved, the greater the number of discourses, values and representations and the greater the potential for conflict. By incorporating notions of procedural justice into communicative action we can begin to develop a praxis of social-planning decision-making, able to incorporate difference and oppositional views and ways of knowing, informed by principles of justice. Through careful organisation and communicative processes, conflict can be sublimated into constructive argument. In this chapter, recommendations are made for procedurally just communicative citizen involvement in the decision-making processes of land-use planning. It is not sufficient, however, simply to invite people to become 'involved' in land-use planning processes that remain inside a modernist, rational, comprehensive framework. We need to create institutional processes that will enable such participation to occur meaningfully.

Empirical patterns of land use are often the results of a complex interplay of actors' diverse cultural perceptions, images, values and mind-sets underlying the activity of local planning policy-making. To divorce analysis of the output from the process amounts to divorcing analysis of action from discourse. Such a divorce misconstrues the nature of planning decision-making. As a result we fail to understand the choices through which people become participants in processes and contribute to decisions.

The ways in which different planners and community groups perceive themselves, their own or others' geographical places and other groups respectively, reveal cultural differences of self- and place-identity, of discourses and values; in other words, situated knowledges. Individuals and groups define and reinterpret place, symbols and practices, mobilising different logics to serve their purposes. As these actors struggle to transform the social relations between them, they produce new 'truths' by which to understand and explain themselves, their practices and their societies.

Planning is an activity laden with the values and perceptions of all the actors involved. Planning is inherently communicative, cultural and political. Theories of planning, to be useful, should therefore be grounded in empirical analysis of what planners and other actors do, be sensitive to the practical situations which planners face and seek to interpret and understand, and be critical of the ethical, socio-cultural and political underpinnings and consequences of action (Forester, 1993a, p. 16).

In this chapter I outline briefly several of what I believe to be key elements of planning praxis: the role of self- and place-identity and recognition, actant-networks and communication. Without communication we have no means of understanding actors'/actants' behaviours and feelings. Haraway (1992) explains that actors are not the same as actants. Actors operate at the level of character and actants at the level of function. Several actors may (but need not) comprise one actant.

Box 14.1 A Western Australian context

Western Australia uses a predominantly British planning system: (the 1928 Town Planning and Development Act, as amended incrementally). The focus is strongly on land-use.

Planning is zoning-ordnanced. Land-use zones are drawn up in the WA Metropolitan Region Scheme as a broad-brush document for the Perth metropolitan region, and by constituent local authorities in their individual town planning scheme.

Planning is rooted in an early 1960s, rational comprehensive style of zoning master plans. There is a dominant development ethos in Western Australia, reflecting the liberal, rational economic outlook of the current state and federal governments. In such an ethos, social planning is invisible. In the buzzword of planning *for* 'community', this means physical and behavioural determinism, creating 'community' through physical means such as zoning and urban design.

Sustainability is economic rather than environmental or social. When Australia hosted the Local Agenda 21 International Conference in 1997, no local authority in Western Australia had a Local Agenda 21 in place at the time.

Despite the Local Government Act, 1995, requiring public consultation, public participation in Western Australia is largely cosmetic. Benefits, if any, tend to accrue to dominant, organised groups.

There has to exist the will on the part of the powerful to move towards building a shared understanding and practice of planning. One of the ways to begin to do so is to incorporate some of the ideas of communicative action and procedural justice into a praxis of procedurally just communicative action, which I develop towards the end of the chapter. I conclude by thinking reflexively on land use planning processes and looking forwards to the possibility of some form of associative democratic structures which may help us to realise collaborative strategy-making and participatory democratic planning in a procedurally just manner.

Identity and recognition

Critical theoretical approaches recognise the contingency of knowledge and understandings of self- and place-identity. Recognition that norms, beliefs, identity and practices are intersubjectively constituted and historically and contextually contingent can enable us to explore different actants' images and identities and how they relate in the planning process. It offers the opportunity to examine the 'knowing from within' of actors and how such knowing is placed within social, moral and political systems.

As Massey (1991, p. 278) writes, 'there are indeed multiple meanings of places, held by different social groups[and] the question of which identity is dominant will be the result of social negotiation and conflict.' There can be no one reading of place. Planners have traditionally sought to 'balance' readings, but have often brought particular mind-sets to the very act of 'balancing', which have, perhaps unconsciously, biased the scales.

Planners may also underestimate the importance of residents' attachment to their local areas and how it comprises a vital component of their social identity. A threat to their physical environment thus becomes a threat to the self. Traditional forms of planning decision-making have tended to convey a message of place as identified and controlled by outsiders (the planners). Plans and policies are loaded with material, ideological and political content, which may perpetuate injustices and do violence to those values, images and identities that have not been traditionally recognised. 'Instead, then, of thinking of places as areas with particular boundaries, they should be imagined as articulated moments in networks of social relations and understandings' (Massey, 1991, p. 280).

In addition to the discourse of planners, we need to pay attention to the ways in which other people verbalise their places in the world, their values and identities. Meanings may become more important than facts in policy deliberation.

In order to understand the behaviour of actors and actants in planning processes, it is important to comprehend their images/identities of who, where, why and how they are. The late 1980s and 1990s have seen a geographical re-emphasis on the importance of space and place and, often, drawing on the work of Michel Foucault, their interconnections with power.

It is now generally accepted that space and place are socially constructed out of human interrelations and interactions. Several authors have rejuvenated

Hagerstrand's concept of space/time geography (e.g. Massey 1993), whilst in debates around identity, space and place also figure prominently. Pred (1984, p. 279), for example, identifies place as involving 'an appropriation and transformation of space and nature that is inseparable from the reproduction and transformation of society in time and space. As such, place is not only what is fleetingly observed on the landscape . . . it also is what takes place ceaselessly'. That which 'takes place ceaselessly' is not simply physical creation and de-creation of landscape, but also socio-cultural mediation of a 'felt sense of the quality of life'. (Pred 1983, p. 58)

Places become repackaged and re-imaged to become attractive to developers and their clients. Place is a saleable commodity. As such it is necessary to remove or decrease the influence of potential obstacles to development (such as public opposition). Governance engages in 'selling' the image of a particular geographically defined place in order to make it attractive to economic entrepreneurs, tourists and/or potential residents. Central to the activities subsumed under selling places is often a deliberate and conscious manipulation of culture in an effort to enhance the appeal and interest of the place. There are conflicts over images and representations of the area by various groups of local citizens, levels of governance, developers and other interest groups.

The rise of the entrepreneurial state has also complicated the picture. In instances where the state is both a major land/property-holder and developer, the state is articulating its own needs and interests. As Howitt (1994, p. 4) comments, 'the channels of ideological power are dominated by developmentalist thinking and values in ways which discredit and marginalise alternative constructions and interpretations of emergent geographies'.

The shaping of place is therefore the outcome of power struggles between various cultures, aspirations, needs and fears within the existing social order. Harvey (1973, p. 31) suggests if that we are to 'evaluate the spatial form of the city, we must, somehow or other, understand its creative meaning as well as its mere physical dimensions'. We need to go beyond identification of conflicts of interest to unpack the cultural differences which influence participants' ways of giving meaning, value and expression to tangibles and intangibles: to look not only at a culture's own interpretation of things, but its interpretation of other cultures and the different aspects of meaning that are formulated through such interpretations. We need to understand how the same sets of signs are read differently by different people and how people make connections between their interpretations of things and the overall ordering process of the planning system. Meaning is a socio-cultural construction which foregrounds representational processes (such as public participation strategies) as hermeneutic and ideological.

It should not be surprising, then, that people react to proposals for urban change in a variety of ways, not all of which are comprehensible to others involved, yet which may well be patterns of action guided by deeply entrenched beliefs, norms and values. Many of these different images and values conflict with and counteract each other.

Other important aspects influencing outcomes include structural power differences between participants and their networks, which may give some people more advantage or control over others; the history of relationships between participants, including previous experiences and prior attitudes and beliefs about each other which participants bring to the discussion; and also the social environment in which participation takes place. All the above serve to influence the type of relationship participants will have with the networks and processes in which they engage with each other.

Land use planning decision-making processes, with their various opportunities for public participation, are thus a series of nodal points, temporary points of fixation in time, at which actants bring together their different representations. Places are therefore articulated moments in networks of social relations and understandings.

In public-participation processes there are often tensions and collisions between different cultural values as conflicting representations are temporally linked in discursive contestation. I take brief snapshots of three key, and related, issues from Australia, to tell stories illustrating these tensions and which link into other themes in the book:

- readings of natural environment and its value
- sustainability
- Aboriginal people and land

In the tone, as well as the content of the stories in Boxes 14.2–14.4, we recognise 'issues, details, relationships and even people' (Forester, 1993b, p. 31) who have been ignored and unappreciated in the past. A planner's view of the world tends to be Euclidean and instrumental. It regards the area in two-dimensional form on a map, geometrically divisible into discrete lots for the provision of housing and urban infrastructure, and as having no value in itself.

Planners' visual geography is presented as being objective; a verifiable truth. However, this composition of view is subjective, determined by cultural (in this instance, professional) practice. The maps and plans which result from the planner's view suggest

> a detachment from the world and a power or control over the world in the hands of the map owner (or reader) which previously was the preserve of God. This is perhaps to speculate, but nevertheless the bird's eye view is a different style of use of the eye to everyday visual experience down on the ground in amongst the houses and streets, farms and fields of day-to-day life.
>
> (Rodaway, 1994, p. 141)

It is to the unpacking of day-to-day life that I now turn.

Actant-networks

In an actant-network, actors and actants in discrete situations become bound into wider sets of relations, which provide forms of identity and the basis for

---

Box 14.2 **Readings of the natural environment and its value, snapshot 1**

Romantic

We need it for our own sense of place: an area of solitude; it provides a break from everyday living.

(Resident)

It's the wildnerness-type concept. You know, you get out in the bush and you just wind down.

(Resident)

Conservation

The site has a lot of conservation value – sand dunes, flora and fauna.

(Resident)

It's got undulating dunes, well vegetated and they are quite unique. I think they should be preserved.

(Resident)

There's an integral conservation value to certain areas . . . flora and faunas; conservation and natural environments have value . . . nature has rights not to be bullied and trampled without its conservation value being assessed.

(Environmental consultant)

Scientific

You have to look at species types and percentages

(Planner)

The scientific emphasis is on preserving a representative sample.

(Planner)

The parabolic dunes at Burns Beach do not qualify as elongated parabolic dunes (length/width ratio of approximately 1.6) and appear to be an example of the more simple crescent form variant, with some possible nesting of a secondary Q2 dune line closer to the point of origin.

(Planning consultant)
(All quotations from Hillier and van Looij, 1997a)

The romantic narrative locates representation of the environment as a distinctly cultural rather than a scientific perspective. It reflects what people value in their 'natural' environments: peace, beauty, birds, flowers and trees and links these with spiritual, mental and physical well-being. It 'explicitly links our biospheric concerns with our other cares . . . [and] offers the possibility of addressing environmental issues "in the round", as material, moral and emotive-aesthetic

(Healey, 1997, p. 185)

Although the conservation-related quotations demonstrate a predominantly aesthetic view of nature, there is also an indication of some element of scientific determination of what should be conserved and why. The language of the scientific statements excludes comprehension by laypeople. There is an overall feeling of the environment as a commodity or resource, but, as Healey (1997, p. 179) asks, 'is it possible to treat biospheric relations in the language of stocks and assets when it is the ecological *relationships* which are critical?

(emphasis in original)

Box 14.3 Sustainability, snapshot 1

On government proposals (Agreements) to log old-growth forests

East Gippsland is a case in point. The recently signed Regional Forest Agreement . . . is underpinned by independent scientific assessment, which builds on about 25 years of reports on land and forest use in the region . . .

The economic fabric of East Gippsland has largely been constructed around the timber industry which is responsible for up to 40 per cent of local employment.

Australia now runs a current account deficit of $2 billion in forest products by importing from countries whose forest management and environmental controls are generally inferior to ours. This is costly to our economy and to the global environment. . . .

By combining the competitive strengths of our timbers, Australia has the potential to lessen its dependence on imported forest products and become a net exporter at a time when global shortages are emerging. . . .

We do not live in a utopia, where we can ignore the commodity values of our forests while using the products from someone else's forest.

(Director, Victorian Association of Forest Industries, in Frazer, 1997, p. A13)

The Federal government has invited woodchipping in areas that were known habitats for threatened species, including the glossy grass skink, powerful owl and spot tail quoll. What they've done is to stitch up woodchipping in East Gippsland for the next 20 years. A lot of species already close to extinction will continue to have their habitat woodchipped.

They've basically let the chippers write the agreement.

(Environment Victoria campaigner, in Winkler and Mitchell, 1997, p. A6)

The deal has not only ensured forests would be a hot issue at the next federal election, but the agreement is calling on forest protesters to take up the fight to protect the animals and species that are going down. East Gippsland is far too important to allow this destruction to continue.

(Wilderness Society spokeswoman, in Winkler and Mitchell, 1997, p. A6)

The forest industrialist above regards forests as a commodity; a source of wealth, at both local (in terms of employment and multiplier effects) and national (export dollars) levels. He also emphasises the importance of a global picture. Trees have extrinsic value, as a resource to be exploited. The environmental spokespeople perceive the intrinsic value of the forests, in their species of fauna and flora. They speak of threat rather than opportunity, using the rhetoric of war and fighting.

Box 14.4 Aboriginal people and the land, snapshot 1

Faced with proposals for urbanisation

A Nyungah elder stood beside Yarkiny (Turtle) Swamp, rubbing his grandson's hair. 'See our ancestors, they're all there around the swamp. See them, telling us to take care. . .

> The whole land is our mother, says an elder. And on a mound above the creek, a gently rising hill becomes the mother's breast, essential for the spring of milk, a stream to be kept clean or the children will suffer. Removal of the breast endangers life and can only be done with full recognition of the consequences. Nyungah ancestors were watching, all around all of us. This is spirit-land.
> (Nyungah Aboriginal resident, in Cunningham, 1994, pp. 7–8)

> A senior planner was asked how he thought 80,000 people would impact on the Swan River downstream from the development. He looked puzzled: 'that's nothing to do with me, you'll have to take that up with another department' . . . Nyungah religious belief and wisdom were as alien to him as the notion that straight rows of brick houses, highway pollution and removal of material bush are not progress. He was offended by the bush: 'I'm just back from Europe. Europeans have made something of their countryside, a lovely green. I don't feel at home in all this,' he confided to a visitor.
> (Cunningham, 1994, pp. 7–8)

> It has taken us a long time to understand and find out about all that is planned and we still do not know if we have covered everything, such as only today did we know that a section of West Swan Road marked in blue meant that it would be widened which may disturb or completely destroy the Grave of our Ancestor, Yagan, whose head was chopped off his body and sent to England. . . . The Nyungah people are of a different culture and understand the land in their own way, not through white man's maps and plans.
> (Excerpts from unpublished letter from the Nyungah Circle of Elders to the Western Australia Minister for Planning, 24 April, 1993)

The stories quoted above highlight the entirely different world in which Aboriginal people live, as compared to the concerns of the planning system. We see the importance of cultural argument against technical argument, the perception of paper, reports, maps and charts, and of the planning system as a whole, as alien. Memory and tradition are keys to beginning to understand Aboriginal attachment to the land. Memory is embodied in identity. There is little objective distinction between space and time.

action. In activities such as public participation in land-use planning decisions, several different actant-networks and non-human intermediaries between actants and networks (such as texts or money) will overlap and align with each other.

Callon (1986) terms the act of an actant constructing a new network(s) by drawing upon actants and intermediaries already in established networks, (e.g. the local authority planning system, residents' associations, etc.), and exerting itself upon others as 'translation'. This process involves the following four stages, although these are not necessarily sequential, or mutually separate:

- incorporation: actors/actants join and are woven into networks;
- interessement: actants exert influence over others via persuasion that their position is the best. Competing alliances are undermined;
- enrolment: actants lock others into their definitions and network so that their behaviour is channelled in the direction desired by the enrolling actant(s);
- mobilisation: the actant now speaks for/represents the others who have become 'redefined' and passive. The representations of interest made by the lead actants are accepted as legitimate by those ostensibly being represented. The represented are reduced to being recipients of action.

Translation is therefore the mechanism by which society takes form. Unpacking these mechanisms enables us to begin to understand some of the power relationships in the land-use planning decision process; an explanation of 'how a few obtain the right to express and to represent the many silent actors of the social and natural worlds they have mobilised' (Callon, 1986, p. 224).

The notion of power is central to the actant-network approach, developing, as it does, Foucauldian ideas of power/knowledge. Action is power-full. 'Those who are powerful are not those who "hold" power but those who are able to enrol, convince and enlist others into associations on terms that allow these initial actors to 'represent' all the others' (Murdoch and Marsden, 1995, p. 372). In so doing, they displace or speak for the others whom they have deprived of a voice by imposing their definitions, images and values upon them. A network is thus composed of representations of beliefs, values, images and identities, of self, others and place.

Actants will utilise whatever resources/intermediaries (including scientific documents, surveys, petitions, etc.) are available to them in order to persuade other actors to their representation or view in the pursuit of their goals. Inevitably, some actors/actants will be able to mobilise a greater quantity of resources than others. In addition, although representers claim to speak for those represented, 'a representation cannot capture all there is to be represented' (Marsden *et al.*, 1993, p. 31). The represented, or non-present, (e.g. nature, people of lower socio-economic status who tend not to participate, those not yet moved into the area, the unborn, etc.) could always 'speak' differently. Translation, if left to its own devices, is seldom equitable or just.

Places are therefore 'shaped' by the representations of actant-networks. They are dynamic; constructed representations by actors/actants at a particular point in time, building upon the remains of previous rounds of representation and struggle. Translation becomes a struggle for discursive hegemony and a means of representing place, remaking it and altering it in the process.

Examination of networks enables us to unpack how domains of inquiry and representation are constructed so that certain representations can be made and others cannot, 'not because of any explicit prohibition or because they are clearly beside the point, but because of ossified perceptions of what the point is; because they do not fit, do not mesh with what points in the established discourse know how to respond to – and because of a reluctance to reconsider those boundaries' (Code, 1995, p. 4). The identification and definition of a planning-policy problem therefore cannot be taken for granted, but depends on its discursive representation.

It is also important to note that the struggle over translation does not take place in a vacuum, but in the context of the existing institutional praxis and actor/actant-spaces of planning. This gives planners and those participants who are comfortable with the discourse of planning a distinct advantage over those who are not.

The role of communication

Communication is vital to processes of self- and place-representation and the acts of planning decision-making. Communication both makes meaning and transmits meaning. In the utterances of communication. 'We are confronted with the actors' own theorising, interpretations, articulations of self and other' (Forester, 1993a, p. 2).

Actants communicate using a range of different discourses; of ideology, of language, of style and so on. Public participation exercises often represent a struggle of discourses between the different communities involved, between the different systems of meanings and frames of reference that are drawn upon, and between the different languages used. By unpacking communication we can begin to recognise the ways in which actant-networks express knowledges and traditions, articulate values and make claims for policy attention.

Actants communicate for specific purposes. They give information, demand information, make claims for policy attention, make proposals or even threats, build knowledge and exchange power. (See Healey and Hillier, 1996, for detailed analysis of the communicative micropolitics of a public meeting.) However, actants from different cultural lifeworlds, using different discourses, may talk *past* or *at* each other rather than with each other. Many actors become intimidated and their voices disempowered. Planners, in particular, may use incomprehensible jargon (plannerspeak), failing to recognise that their own professional terms and definitions deny themselves, as well as other participants, the opportunity to ensure that their stories are respectfully understood.

Planners are experts. As Giddens (1994a, p. 84) points out, the difference between experts and laypersons is often a difference in power, but essentially it is an imbalance in skills and/or information that makes one an 'authority' over the other. Laypersons' knowledge, as we have seen, embodies tradition and cultural values; it is local and de-centred. Planners' expertise, on the other hand, is disembedded, 'evacuating' (Giddens, 1994a, p. 85) the traditional content of local contexts, and based on impersonal principles which can be set out without regard to context; a coded knowledge that professionals are at pains to protect.

Planners traditionally believe themselves to be neutral, rational experts, offering objective and balanced appraisals rather than making value judgements. Yet planners must inevitably bring their own values into their work, making judgements as to the good versus the right; what it is important, which interests should carry how much weight, what is possible to achieve and so on. Planners and governance reserve the ultimate power to define, redefine, organise and reorganise space into a place of their choosing.

Box 14.5 Readings of the natural environment and its value, snapshot 2

Communication and intermediaries

Press releases

> Reject the unacceptable development proposal and act to protect the bushland as an A-class reserve . . . this is one of the few places on the metropolitan coastline where we can establish a major conservation reserve.
>
> (Resident action group press release, *Wanneroo Times*, 12 December 1995

Petitions and form letters

> Form letters and petitions helped those who were short of time and/or didn't feel qualified.
>
> (Resident)

Letters to the press

> For our children whom we love and cherish more than anything in the world, please help save the Burns Mindarie Dunes.
>
> (Resident, in Wright, 1995)

> The public should be permitted to deal with fact, not emotion . . . the area contains no rare or endangered species.
>
> (Developer, in Clarke, 1995, p.11)

The actant-network of local residents has the resources (financial, expertise, time, personnel) and dedication to attempt to enrol others to its representation of the threatened dune area. The developer has also been forced to use the press defensively and as a means of trying to persuade others of his case.

Box 14.6 Sustainability, snapshot 2

Logging old-growth forests: communication and intermediaries

Regional Forest Agreements

The Prime Minister, Mr John Howard, said the agreement would generate investment worth $150 million and create 400 jobs by setting out forest usage in East
Gippsland for the next 20 years and providing resource security to the timber
industry.

(Winkler and Mitchell, 1997, p. A6)

Press releases

Many groups, such as the Wilderness Society, believe there is no longer any place for
native forest logging. Just as whaling is a socially unacceptable way to provide jobs
and just as there are alternatives to whale oil, there are alternatives to felling the
leviathans of our ancient forests.

(Wilderness Society campaigner, in Redwood, 1997, p. A13)

The nationally significant rainforest area of Goolengook being cleared in East
Gippsland is estimated to yield 75 per cent woodchips. It will provide five jobs for
about four weeks.

(Wilderness Society campaigner, in Redwood, 1997, p. A13)

In the Gippsland region, the logging industry has an annual turnover of $53 million.
This accounts for less than 13 per cent of the region's income. Tourism alone
accounts for $134 million and is a rapidly growing industry, with about 2.5 indirect
jobs created for every one directly involved in tourism.

(Wilderness Society campaigner, in Redwood, 1997, p. A13)

Demonstrations and protests

The group has to decide quickly what their response will be . . . By 5am their
response is already made. It is shockingly cold . . . The sun won't be up for a
few hours but Pete is up the tripod, which was erected at 3am. Now the road
is blocked. . . . the loggers turn up. They skid to a halt in front of the
bridge. One truck narrowly avoids collecting Pete and almost lands in the creek.
The loggers groan and laugh as they greet the protesters. They call for the
police.

(Trioli, 1997, p. A6)

The actant-networks of forest protectors are using intermediaries of high profile
protests and national media coverage in attempts to enrol as many Australians as
possible into their representation. The environmental spokesperson counters the discourse of the government press release by communicating using similar language
and factual style to persuade her audience of her case.

Box 14.7 Aboriginal people and land, snapshot 2

Intermediaries

Writing letters and talking to planners

We write letters after letters and we get no response.
We've tried time and again. We've left the gates open. We've tried to coax them to come and listen. We've talked and talked and talked.

(Bropho, personal communication, 1996)

Talking to developers

We met, us the blacks and the people. We sat down and we talked. Just two groups and we came up with a solution.

(Bropho, personal communication, 1996)

We set down this idea of we'll meet only with developers, which we feel is the right way to do it.

(Bropho, personal communication, 1996)

Written agreements

. . . written this letter which is going direct to the developer, saying meet with us direct, meet with us in good faith and come to an agreement with us. . . . We've seen already one who've said on the phone they're prepared to meet under those terms and come up with a written legal agrement.

(Jefferies, personal communication, 1996)

Using legislation

One other way we're looking at now is we'll meet under the Mabo legislation or the land titles legislation.

(Bropho, personal communication, 1996)

The Nyungah Aboriginal people have lost faith in dealing with planning officers and are turning to communicating directly with developers as a means of enrolling them into their representation of the local area and its spiritual significance. The Aboriginals are also using the Federal Native Title (Mabo) legislation (which accepts Aboriginal people as custodians of non-freehold Crown land) as an intermediary to further their argument. It is unfortunate that the Nyungahs feel that they cannot talk to governance, although if the following statements are typical their position is understandable

If you bring in minority groups [i.e. Aboriginals], they focus on their own need, have their own agenda, alienate the institutions and the majority of the community and take over.

(Local authority councillor)

Involving them reduces our credibility as a group with the departments and instrumentalities.

(Local authority councillor; Hillier, 1998)

Through communication, therefore, planners and other actants often seek to enrol other actors into their representations. Their goal is mobilisation; acceptance of their representations as legitimate. Public participation programmes are often utilised as the means of persuasion, involving intermediaries such as press releases, letters to the media, petitions and so on. Through communicative analysis we can recognise 'the important dimensions which need to be understood if government action is to break away from its damaging effect on the public sphere and engage in real discussion with citizens about the future of their areas' (Healey and Hillier, 1996, p. 182).

Communication enables us to identify claims about value, claims about what actors are worried about, want to gain, are afraid of, wish to protect, or care about enough to put on the table for discussion (Forester, 1996). Representations are complex, dynamic, multifaceted, contested and unfolding rather than the neat singular, linear constructions preferred by the Cartesian view of the planning system.

Communicative action

The concept of communicative action (developed from Habermas, 1984, 1987) offers a methodology by which policy decisions can be made through people reaching a position of mutual understanding by inter-subjective reasoning and discussion. A communicatively active process enables actants to negotiate decisions whereby everyone is allowed to have a voice and that voice is listened to with respect for its opinion. Actants thus begin to understand the interests, perceptions and constraints on their co-negotiators, and may begin to revise their opinions and expectations accordingly. If professional experts and citizens attempt to understand the discourses, meanings and lifeworlds of the other 'sides', a creative form of decision-making could evolve in which areas of common ground or overlapping agreement emerge.

One of the advantages of decision-making through communicative action is that no decisions should be forced on the participants. Agreements should be voluntary, based on reasoning and good argument. The legitimacy of the final agreement thus increases as all sides will have contributed to the outcome and own a share in its success. Collective decisions tend to be accepted more easily and opposed less than unilateral judgements.

A further advantage of communicative action should be that planning becomes open and transparent. Expert, professional technicalities and jargon would be explained in a sharing of information and demystification of scientific argumentation. The debate not only incorporates valuable input from people who would otherwise be excluded from technical discussion, but also becomes opened up to a wider number of policy alternatives. Discursive decision-making should thus broaden the concept of what is a 'valid' argument from having a merely rational technical basis to incorporating a 'wide understanding of what we know and how we know it, rooted as much in practical sense as in formalised knowledge' (Healey, 1992, p.153).

Is reasoned agreement between all participants ever possible? Actants in public participation programmes often hold diametrically opposed views. Can we do justice to all values, images and identities and still negotiate consensus? I suggest not, in the traditional meaning of the term 'consensus'. We need to include in our notion of consensus, the possibility that differently formulated identities, values, claims and arguments may find any common links extremely precarious, and that there may well be substantive and even intractable disagreements over basic issues. I prefer to adopt Love's (1995, p. 62) version of con-sensus, spelled with a hyphen, carrying a meaning of 'feeling or sensing together', not necessarily implying agreement, but involving a respect for the views of others and an attempt to understand them.

Con-sensus enables local actants in participatory decision-making programmes to resist mobilisation and the colonisation of their lifeworlds by systemic power structures and to replace them with a new consciousness of multiplicity. We need to reconstitute the Habermasian concept of the public sphere as one that allows for the creation of local, autonomous public spheres as appropriate, to which all members of the public have access as relevant and which guarantees respect of differences, freedom of expression and the right to criticism.

Such public spheres would recognise a multiplicity of different meanings, ways of knowing and forms of expression that traditionally have had no or little legitimacy in the planning process. How does such recognition of and respect for difference between participants fit in with a Habermasian notion of communicative action? Habermas (1992) himself has now begun to address the problem of difference, arguing that universalisability of needs recognition, etc. does not imply that differences in concrete forms of life must be levelled, but that participants should be afforded recognition as unique individuals. The unity engendered by communicative action does not eliminate the difference between individuals, but confirms it.

It is through communication, through dialogue with others, that people can understand their differences and reach 'a contested but negotiable practical understanding' (Shotter, 1993, p. 116 of different non-assimilable ways of being and ways of knowing. If communicative action, then, need not level out differences between individuals, we need to develop a form of its operationalisation grounded in a transparently just procedure.

Procedural justice

As we begin to recognise the validity of differences in actants' modes of thought, identities, images and values, and to add difference to a Habermasian communicative action framework, we now face the problem of how to treat all actants fairly. 'Can we negotiate our way through interest conflicts by sensitive bargaining and respect for individual preferences, whatever they are, or is the range of claims for attention now so great and so interconnected that it defeats our negotiative capacity?' (Healey, 1997, p. 246). Can we create, in effect, Giddens' dialogic democracy (1994b, p. 115)?

As I have indicated above, I am gradually shifting from a traditional form of consensus to the idea of a con-sensus reached through a procedurally just con-versation between participants, which everyone is prepared to deem reasonable and fair. I now turn to the concept of procedural justice as offering us guidelines for ensuring that all participants are treated, not necessarily equally, but fairly or justly in the decision-making process.

The intuitive idea behind procedural justice is very simple: do unto others as you would have others do to you. As Lind and Tyler (1988) indicate, the politi-cal arena (including town planning policy decisions) is an important site for implementing procedural justice. It is important to determine not only that the outcome is fair, but also the extent to which the procedure resulting in the out-come is fair to all those affected.

At the heart of the notion of procedural justice is the idea that, as far as possi-ble, people should be asked to participate in and consent to decisions that affect their interests, particularly in cases where contemplated actions are anticipated to produce adverse effects. Research (Lind and Tyler, 1988, and others) has found that decisions are more likely to be accepted by those affected when the procedure used to generate the decision allows their perceived fair participation. Fair partic-ipation comprises having the ability to express one's opinions and tell one's stories (voice), being listened to with respect, having access to adequate information, being able to question others, having some degree of control over the decision-making procedure and resultant outcome, demonstrating that decisions are made impartially and receiving good feedback. In addition, moral traits such as perceived trustworthiness and honesty, and mutual respect are acknowledged to be impor-tant in generating fair, co-operative decisions.

I believe that if procedurally just processes of participatory decision-making are followed in local planning practice, putting together components of procedural justice with those of communicative action, then we may well move towards the realisation of a situation at a local level wherein the planning decision system becomes structured not only by the planners' systemic needs of 'control' (i.e. statutory procedures and legislation), but also by the values and needs arising from the lifeworlds of other participants.

Procedurally just communicative action

A participatory, communicative approach to decision-making is never easy. The greater the number of actors involved, the greater the number of discourses, rep-resentations and identities and the greater the potential for conflict. Procedural justice becomes an important aspect of decision-making if actors are to be satisfied with the decision process and outcomes.

I believe that procedurally just communicative action points the way forward. Planners cannot afford to be 'outside', neutrally detached from the world. They must be 'actors in the world' (Innes, 1995, p. 184), and leave behind many of their traditional instrumentalist assumptions about knowledge and method. By incorporating notions of procedural justice into communicative action we can begin to develop a praxis of social planning decision-making able to incorporate difference and oppositional views and ways of knowing and which is informed by principles of justice. Planning decision-making processes are re-sited within the plurality of the social. Process elements of procedurally just communicative action may also include affirmative or positive action as appropriate, the use of a facilitator, the incorporation of a 'cooling-off' period and so on, in addition to those aspects of procedural justice listed above.

Reflexions and forwards

It is not sufficient simply to invite people to become involved in public participation exercises for land-use planning processes which remain inside a modernist, rational, comprehensive framework. We need to create institutional processes that will enable such participation to occur meaningfully. Through careful organisation and procedurally just communicative processes, conflict can be sublimated into constructive argument.

As Healey (1996, p. 213) indicates, 'the key to effective institutional design includes fostering an interactive and collaborative capacity, building the social capital of trust and the intellectual capital of understanding, even across deep divides and tensions'. She points out the importance of three key features:

- designing arenas for communication and collaboration which give access to all those who have a stake in an issue;
- finding ways of conducting discussion and shifting decisional power as close as possible to those who will experience, and live with the consequences of strategic choices;
- fostering styles of discussion which allow the different points of view of diverse stakeholders to be opened up and explored.

(Healey, 1996, p. 213)

The difficulties of implementing the above, even given a commitment to 'rules' of procedural justice, should not be underestimated. Yet both Healey and I believe that such rules are necessary, as they help to 'force the "powerful" to pay attention' to all other participants. In this spirit I offer some recommendations towards procedurally just communicative action (see Box 14.8).

Citizens have now come to expect, and therefore demand, a voice in local planning decision-making. The rise of interest groups such as residents' and ratepayers' associations, Friends of . . . and other voluntary associations will not be turned back. I believe that the concept of associative democracy represents a valuable means of linking public sector policy-formation and implementation with the needs and interests expressed in the non-public sector. It may well offer us the structures that Healey and others seek and help us to realise collaborative strategy-making and participatory democracy in a procedurally just manner.

Associative democracy has developed theoretically from the Habermasian notion of communicative action and relates to current pragmatic theories of planning praxis. What might associative democracy strategies imply? I have identified

Box 14.8 Recommendations towards procedurally just communicative action

Land-use planning decision-making should follow strategies of collaboration, including seeing the issue as one to be solved in a mutual way.

Land-use planning decision-making processes should distinguish the needs, values, interests and positions of all participants in relation to the major issues in question.

Land-use planning policy decision-making processes should focus on the perceptions, cognitions, needs, fears and goals of each participant and allow for the exchange of clarifications, acknowledgements and assurances.

Land-use planning policy decision-making requires clear and honest communication.

Citizen involvement in land-use planning policy decision-making can usefully be facilitated by a skilled third party who enhances motivation, improves communication and advocates as socially and ethically appropriate.

Citizen involvement in land-use planning policy decision-making should take an approach of face-to-face interaction under norms of mutual respect, shared exploration and commitment to resolution.

Interaction between participants in citizen involvement in land-use planning policy decision-making should incorporate the qualities of productive interpersonal negotiation, including, open and accurate representation, recognition of interpersonal and inter-group diversity, racial, ethnic and gender equality, integration of all participants' knowledges and skills, sensitivity to cultural differences and power imbalances and persistence and discipline to attain mutually acceptable outcomes.

Citizen involvement in land-use planning policy decision-making should incorporate rehabilitation processes at an early stage if appropriate for both officers of governance and communities.

Land-use planning policy decision-making processes should take an approach of shared planning and decision-making between governance and communities. This process should specifically:

- develop an agreed and ethnically sensitive public participation process with the communities;
- clarify the roles and responsibilities of different government departments involved;
- clarify the roles and responsibilities of citizens involved;
- ensure community involvement in the early stages of 'problem' identification;
- widely promote opportunities to be involved;
- select the public participation methods utilised from the range of appropriate techniques;
- clearly communicate the process and what is open for negotiation;
- clearly communicate who will be the final decision-maker;
- involve a broad range of interests, both community and governance;
- ensure full, unbiased and comprehensible information is easily accessible using appropriate means of communication (e.g. local ethnic radio) and give assistance with understanding of information as required (e.g. by meeting, letter, etc.);
- raise community awareness and understanding of planning issues;
- broaden the range of acceptable means of community voice (e.g. submissions written in languages other than English, comments made by telephone, on cassette tape, etc.);
- listen to all communities respectfully; ensure independent decision-making;
- let participants know how their views have been incorporated (or not) into decision-making;
- ensure action on the final decision by the relevant agency.

(Hillier, 1997)

above how representation and the persuasion/influence of actant-networks play important roles in planning decision-making. Associative democracy, through the politics of associations of actant-networks, may open up ways through which the exercise of power and persuasion can meet democratic norms, regulating the more wealthy and larger interests and enhancing the voices of the less articulate and smaller interests.

Associative democracy thus builds on already existing actant-networks and the potential for actant-networks to form in response to planning issues. It also theoretically enables those people who do not tend to join networks (especially the poor – see Hillier and van Looij, 1997b) to form their own. Associative democracy therefore recommends promoting the organised representation of presently excluded cultures and interests. Moreover, it encourages the organised to be more other-regarding in their actions and recommends a more direct and formal governance role for groups. This last point is important. There is a need for new institutions if associative democracy is to succeed. Political and socio-economic strategies alone are unlikely to serve the interests of the disadvantaged. There will be a need for judicial strategies. 'The state must basically perform the role of a partner, enabling other actors-partners to perform their own roles; and in some cases it might even have to create these roles' (Amin and Hausner, 1997, p. 18). None of this can be accomplished without using political power to redistribute resources and to underwrite and subsidise (financially, and with time, personnel, etc.) associational activities.

We need to design arenas for communication and collaboration in an associatively democratic manner; finding ways of 'conducting discussion and shifting decisional power as close as possible to those who will experience, and "live with" the consequences of strategic choices [and] fostering styles of discussion which allow the different points of view of diverse stakeholders to be opened up and explored' (Healey, 1996b, p. 213). Procedurally just communicative action is one step towards these goals.

15   Changing cultures

Clara H. Greed

Cultural perspectives

Many of the contributors to this book have commented upon the importance of cultural perceptions of 'reality', as variously held by the planners and the planned, in seeking to make sense of the planning process and have noted that 'the planners' so often make professional decisions that are at odds with what 'the planned' consider needs to be done. Nevertheless, there have been genuine attempts, particularly at the European Union and international levels, to accept the issues of sustainability, cultural sensitivity, accessibility, health, and equality, when 'doing social town planning'. As Hillier comments, in chapter 14, though, when it comes to communication with minority and disadvantaged groups the process of social planning is hindered by lack of mutually understood communication, by people 'talking past each other'.

The construction industry context

This chapter returns to the problem of 'culture' (with particular illustrative reference to gender issues), in respect of the nature of professional subcultures within the construction industry, thus adding another key component to the debate. Much of this book has centred around the world of town planning and what is wrong with it. But 'planning' in the sense of urban policy-making and management takes place under a variety of guises, separately from, and often at a higher level of decision-making than that specific set of activities we call 'statutory town planning' – not least through the activities of other construction professions that have a substantial role in shaping the built environment but, arguably, appear low on social awareness.

The main part of the chapter identifies potential change agents that might transform the nature of professional decision-making and transcend the physical/social dualism inherent in town planning. This chapter draws, in part, upon material from my ESRC research, 'Social Integration and Exclusion in Professional Subcultures in Construction', undertaken in 1997, which extended my previous work on the built environment professions, such as studies of women in general-practice surveying and town planning (Greed, 1991, 1994), into the wider world of construction industry (Greed, 1997b, 1999, forthcoming).

As can be seen from Table 15.1, of the 1.3 million people in the construction industry fewer than 6 per cent of those at professional level are women, whilst ethnic minorities make up around 4 per cent, and the disabled constitute 0.3 per cent, of construction professionals (Ismail, 1998). As can be seen from the last column, all town planners also (coincidentally) only constitute 6 per cent of the membership of the built environment professions (because there are so many more engineers and surveyors). The situation is a little better in professional education, which may bode well for the future of the profession (see Table 15.2). Yet the power of the planners is considerable (Greed, 1994: 21), because as part of a wider local government system they have nationwide regulatory and management powers over the built environment. But the surrounding influence of the other built environment professions and the wider construction industry, of which town planning is officially one small part (CISC, 1994) cannot be underestimated either.

The construction industry is not on a separate planet (although sometimes it seems so); it does not exist in a technological realm separate from the wider society. Indeed, its economic future depends upon being in tune with society's needs and demands. It is doing 'town planning' in all but name, by default. The influence of construction professionals goes way beyond individual building projects: they are among the 'great and the good' being called upon to shape the built environment, at macro- and micro-levels, in a variety of capacities, posts and committees. For example, civil engineers may be heads of town planning departments, transportation planners, and planning inspectors. Construction managers and building surveyors are to be found in the highest levels of academia, including sitting on RAE (Research Assessment Exercise) committees and a range of policy-making governmental committees on strategic urban policy issues.

Since women and other minorities are under-represented in the construction professions, when members are sought for decision-making committees and boards, candidates are chosen from a predominantly male pool. So 'women's issues' are often, by default, dealt with by all-male committees. For example, many BSI committees still predominantly consist of – often elderly – male engineers. The BS 6465 Code of Practice for Sanitary Appliances committee has only relatively recently welcomed its first woman member (BSI, 1995). It is no coincidence that for many years this Code of Practice has specified almost twice the provision of public conveniences for men as against the provision for women (Greed, 1996c), surely an access and equality issue. But 'public lavatories' come under 'building' not 'planning' regulations, as, in effect, do many other social town planning factors, as discussed by Linda Davies in chapter 6.

Although construction professionals emphasise the importance of public service, there is considerable dissatisfaction within the community at the end product in terms of individual building design, inconvenience of location, layout and overall city form, and wider social and access issues. There is concern at the continuing under-representation of minority groups and interests in the development process (Rhys Jones *et al.*, 1996; Druker *et al.*, 1996). This factor undoubtedly contributes to the nature of professional decision-making, and creates an unsatisfactory end product. Clearly, an individual professional's life experience will affect

Table 15.1 Membership of the construction professions, 1996

Body	Student members			Full members			Total membership			% of whole sector
	Total	Female	%	Total	Female	%	Total	Female	%	
Royal Town Planning Institute[a]	2,196	957	43	13,698	3,025	22	17,337	3,972	23	6[b]
Royal Institution of Chartered Surveyors	8,193[c]	1,267	15.5	71,865	4,886	7	92,772	8,062	8.7	35
Institution of Structural Engineers	4,358	586	13.4	10,114	137	1.3	21,636	951	4.4	5
Institution of Civil Engineers	8,353	978	11.7	42,658	767	1.8	79,480	3,425	4.3	26
Chartered Institute of Building	9,859	620	6.2	10,244	94	0.9	33,143	903	2.7	5
Architects' and Surveyors' Institute	—	—	—	—	—	—	5,046	130	2.5	2
Royal Institute of British Architects	3,500	[d]	31	22,670	1,819	8	32,000	[d]	12	11
Association of Building Engineers	327	34	10.4	2,292	39	1.7	4,577	104	2.3	2
Chartered Institute of Building Services Engineers	2,196	116	5.28	6,275	66	1.05	15,264	319	2	3
Chartered Institute of Housing	4,190	2,654	63	8,258	3,465	41	13,490	6,375	47	4
Architects' Registration Board, ARCUK	—	—	—	(UK only)			25,153	2,892	11.5	ARB
British Institute of Architectural Technology (total membership only)	Of 1.3 million in construction, 15% are professional. Of these 6% are women: 0.9 of total			Approx. total for all: 19,500 12,000 6			5,495	182	3.3	2

Notes
a Figures as at end of 1996 gathered in 1997.
b c.g. planners comprise 6 per cent of all construction professionals.
c Changes in total compared with previous figures may be accounted for by revisions in body's categories, etc., but check the overall proportions. In some cases intermediate, honorary, graduate and licentiate categories make up the remainders of the totals.
d RIBA figures are approximate, owing to the new format in which figures for education statistics and membership are presented.

Table 15.2 Undergraduate student survey of characteristics of first-year entrants

Factors		All students			Professional specialisms						
		All UK	Male	Female	Architect	Civil engineer	Building construction manager	Quantity surveyor	Building services engineer	Town planner	Building surveyor
Gender											
1995/6	Male	82	—	—	75	86	93	87	75	68	85
1994/5	Male	84	—	—	75	87	91	87	92	65	94
1995/6	Female	18	—	—	25	14	7	13	25	32	15
1994/5	Female	16	—	—	25	13	9	13	8	35	6
Age											
1995/6	Under 20	64	81	19	66	68	44	64	29	74	54
1994/5	Under 20	75	82	18	81	81	70	67	60	87	47
1995/6	Over 20	36	83	17	34	34	56	36	71	26	46
1994/5	Over 20	25	87	13	19	19	30	33	40	13	53
Ethnicity											
1995/6	White	88	82	18	83	86	91	92	71	90	94
1994/5	White	90	85	15	88	90	90	93	86	88	91
1995/6	Other	12	79	21	17	14	9	8	29	10	6
1994/5	Other	10	79	21	12	10	10	6	14	12	9

Source: Kirk Walker (1997).
Note
All figures are percentages.

his or her judgement of priorities and appropriate policies when undertaking urban decision-making. Many people and professions are involved in the development process, all adding their bit to the design and construction of the built environment, as indicated in Ann de Graft-Johnson's chapter 8 in respect of architecture and planning.

Professional subcultures

Professional decision-making is not socially neutral: it is influenced by an individual's perception of 'reality' – how he (and it is usually 'he' in the world of construction) sees the world and imagines society to be, as already commented upon by Huw Thomas in respect of planners' policy priorities (see chapter 2). I have developed this concept in detail in earlier work on the surveying profession (Greed, 1991, pp. 5–6). It is helpful to see the construction idustry as possessing its own subcultural values and world view. 'Subculture' is taken to mean the cultural traits, beliefs, and lifestyle peculiar to the construction tribe. One of the most important factors seems to be the need for a person to fit in to the subculture. It is argued that the values and attitudes held by its members have a major influence on their professional decision-making, and therefore ultimately influence the nature of what is built. Gender is, of course, a major consideration in the world of construction, as is ethnicity. The need for an identification with the values of the subculture would seem to block out the entrance of both people and alternative ideas that are seen as 'different' or 'unsettling', but which may, in fact, be more reflective of the needs and composition of the wider society.

The concept of 'closure' in the relation to the power of various subculture groups to control who is included or excluded is a key factor in understanding the composition of the professions (as discussed by Parkin, 1979, pp. 89–90, and developed by Weber, 1964, pp. 141–52, 236; see also Greed, 1994, p. 25). This is worked out on a day to day basis at an interpersonal level, with some people being made to feel awkward, unwelcome and 'wrong' and others being welcomed into the subculture, and made to feel comfortable and part of the team. Thus encouraged to progress to senior posts and so to decision-making levels within the subculture, these people are therefore in a position to shape the built environment.

The construction industry is typified by a narrower perspective than the more liberal town planning profession. As to professional subcultures, the emphasis upon specific site development tends to create a tunnel vision which does not allow much space for wider strategic and holistic approaches to urban policy issues or to non-site-specific social factors. Ironically, the industry wants to attract more women, but this is more the result of 'man'power shortages than equal opportunities considerations. The industry is beginning to show great interest in the question of 'how to get more women and other minorities into construction' (CIB, 1994, 1996). Whilst it wants 'more women and ethnic minorities' there seems to be little concern about the likely changes that increased minority representation would bring and demand. (There are parallels with growth of the planning profession twenty years ago.) Clearly 'more' minorities does not necessarily mean 'better' in respect of policy priorities and human resource management, unless fundamental cultural and organisational change also occurs within the profession (Greed, 1988).

Creating change

Contemplating change

A key question arising from this book so far is, 'how can change be generated within professional subcultural groups and transmitted on to the built environment?'. This issue will be explored with reference to my current research as a starting point, cross-referenced to the key points and insights suggested by contributors to this book in their respective fields. The rest of this concluding chapter contemplates the possibilities for change to create 'better planning', in which the social/physical dualism is exorcised or transcended. It is argued that in order to change the built environment, one first needs to change the worldview (that is, the culture, values, perceptions of reality and hearts and minds) of the decision-makers who make the urban policy which shapes our towns and cities. This may involve, for example, changing the social composition of the planning profession even more, so that it is more representative of society as a whole. It is vital to alter the process and personnel involved in the 'reproduction over space of social relations' (Massey, 1984, p. 16). In what often appears to be a fortress-like setting, in which outsiders and minority individuals seeking to bring change are either 'socialised' to conform, marginalised or ejected, this is no easy task.

Conceptualising change: critical mass?

Conceptually, it is important to identify potential agents of change, and to 'map' pathways of change. Concepts which inform the investigation of change include the ever-popular (but somewhat questionable) critical mass theory as to 'how many people are needed to change an organisational culture' (cf. Morley, 1994, p. 195, who refers to Bagilhole's work (1993); and see subsequent work by Bagilhole *et al.*, 1996, on women in construction). Kanter (1977) suggests that between 15 and 20 per cent of an organisation must be composed of minority individuals in order for its culture to change whilst Gale (1995) suggests that 35 per cent is necessary in the construction industry. Respondents have suggested to me that the percentage should be much higher.

Ironically, town planning has one of the highest percentages of women in any of the built environment chartered professional bodies (except for housing) (see Table 15.1) but little change has been manifested, in spite of the constant efforts and writings of 'women and planning' groups. Originally, in physics, from which the theory is derived, critical mass was defined as an amount (like the minimum size of snowball that holds together without melting), that would trigger a chain reaction, not as a percentage. Only 20 pounds of uranium 235 was needed to create critical mass in an atomic bomb weighing 9,000 pounds, which is 0.2 per cent of the total matter (Larsen, 1958, pp. 35, 50, 55, and 73) – comparable to the percentage of minority individuals in some branches of construction, or the numbers of women in some district planning offices!

An increase in minority representation does not necessarily make any difference. It has often been argued that in order to effect change and to make social planning policy more reflective of society there is a need for the composition of the professional bodies, from which the decision-makers are drawn, to be more reflective and representative of society as a whole. But this does not necessarily follow. Much depends upon the personal perspective, ideological base (if any), self-interest, background and education of the minority individuals in question. (Compare the experience of the Swedish Parliament in Stark, 1997, with the New Labour influx of women members in 1997 to the British Parliament). In other words, changing the culture is more important than changing the composition, and quality may be more important than quantity. A small, well-organised group or one or two totally dedicated, charismatic individuals may prove more effective than hundreds of new minority individuals who appear confused and unsure of their role at a personal level, and who have such low expectations and high tolerance levels, that they may declare, 'I simply don't know what the problem is.' They are thus easily socialised into the mainstream.

There are a variety of theories that deal with such group dynamics, with the role of individuals and the significance of networks that ensure either social inclusion or exclusion for 'new' or 'different' individuals (some already referred to in Jean Hillier's chapter 14). Actor network theory, which asks 'what are the networks by which social power is maintained, or the pathways through which social change might be brought about?' is of particular interest; cf. Callon *et al.*, 1986; Murdock, 1997, as is (to mix metaphors) the related role of 'prime movers' in detonating critical mass explosion (Kanter, 1983, p. 296). Also not to be dismissed, are the extensive range of 'New Age' ideas and theories, which delve into more 'spiritual' realms for ideas as to how to create change – the Celestine series is a best selling example (Redfield, 1994). Such work has proved popular among some women's groups seeking social change, such as the Women's Communications Centre, who undertake a massive national opinion survey among women, published as *Values and Visions: What Women Want* (WCC, 1996). When all material, physical, and statutory means have failed, it is necessary to 'ask the Universe': in other words, to pray for change.

Thankfully, it has been found from the research, and from my own experience in the worlds of surveying and town planning, that a resilient and influential minority of women do exist within the industry and *are* likely to hold alternative viewpoints. This group is likely to increase as the present young cohorts of women professionals who might contribute to the build-up of critical mass grow older and become more 'cynical'. But I still hesitate to use the phrase 'critical mass', although I believe there is some truth in the concept that once a certain proportion of women is achieved within the industry the culture will shift – not necessarily for the better, but perhaps in another, unexpected way. In parallel with the situation in physics, it is easier to create an uncontrolled than a controlled chain reaction resulting from detonation, particularly when the setting is unstable (Larsen, 1958, p. 50). 'Critical mass', though is one of the most frequently used terms in the industry when discussing equal opportunities. It fits well with the scientific and quantitative bent of the construction subculture (Larsen, 1956), but is highly optimistic and over-simplistic if used as a predictive social concept without considering qualitative, cultural forces too.

Change agents

Generic or gendered change?

In this section key change agents that might bring about the desired changes will be identified and discussed. Many of these have little to do with statutory town planning, and do not subscribe to the spatial fetishism of land-use plannning, or the 'salvation by bricks' mentality towards solving social problems; by their actions, they are achieving 'social town planning'. Two main types of change agent may be identified, 'top-down' and 'bottom-up'. The first group 'top-down', includes governmental, international, nationwide and professional bodies, and other official agents, many of whom, for a variety of reasons, are concerned with 'generic' change rather than with specifically social issues. Nevertheless, aspects of their work may, incidentally, benefit minorities. The second group, 'top-down', includes a range of minority, grass-roots, community, and single-issue groups, most of whom are concerned with specific minority issues, including 'gendered' considerations. A word of warning in respect of both types of group: not all change is for the good, there are both reactionary and progressive change agents at work.

Top-down change agents

At the international level, a range of top-down change agents, such as the UN, OECD, WHO and the EU itself, have been identified, particularly in chapters 9, 10, and 11. All of these have considerable influence in terms of promoting both 'ideas' and 'procedures' for developing social town planning. Overarching global movements which reflect the *Zeitgeist* of the globe, such as 'the green movement' and more specifically 'planning for sustainability' have been very influential in challenging and reshaping British town planning. In addition, more diffuse, but undoubtedly powerful, cultural movements have been at work, and, as Franco Bianchini indicates in chapter 12, they have made their mark on British town planning. The growth of mass tourism, as discussed by Davidson and Maitland in chapter 13, has created a global culture, enabling the rapid transmission of ideas relating to popular culture, heritage, leisure and public art among competing tourist cities in different countries. This change is reflected in Britain in the renaming of the Department of National Heritage as the Department of Culture, Media and Sport by the incoming Labour government in 1997.

Within the British construction industry and related built environment professions, top-down change agents have pursued more pedestrian, and less glamorous objectives. There are demands for change from a range of mainstream sources. Government initiatives such as the Foresight programme (Department of Trade and Industry) and Partners in Technology (DETR); working groups, including CISC (1994) and the Latham Committee (CIB, 1996); and a range of research projects, have all highlighted the need for cultural change within construction. Reasons for change variously include 'the business case'; increased efficiency of industrial processes within the industry; health and safety considerations; greater competitiveness; recruitment crises; European harmonisation; down-sizing; multi-tasking; greater flexibility; environmental, economic and social sustainability; improved human resource management; qualification rationalisation; educational reform; urban renewal; urban regeneration and the need for creating a climate of technological innovation; economic prosperity and plain old Progress.

Town planning nowadays has been swept up in the demands for cost cutting, speedier planning, greater efficiency, more privatisation, compulsory competitive tender, and entrepreneurialism, just like everyone else. Within this climate, non-profit-making social considerations exist in an uneasy limbo. Many of the above initiatives adopt a generic approach, not 'disaggregating' the needs of different groups upon the basis of gender, race, age or other social differences. It seems that many minority initiatives, groups and key individuals remain 'invisible' to mainstream committees, outside professional networks and terms of reference when it comes to looking for new members or seeking advice on equal opportunity policies. Groups such as Planning Aid for the London and London Regeneration Network (LRN, 1997), *inter alia*, have sought to monitor quite who is asked on to key committees. Unsurprisingly, they have found that many key ad hoc 'social planning' agencies, such as regional committees, urban regeneration boards and housing steering groups have remained white and male in composition, in spite of the influx of so many new minority professionals, particularly in London.

Likewise, higher education is having to show greater productivity, rationalisation and efficiency, accompanied by a spate of modularisation, standardisation, semesterisation, all within a strongly managerialist and, arguably, anti-academic ethos. In spite of all the activity and enthusiasm of minority groups, there is very little space left any more for subjects such as equal opportunities, or even urban sociology. According to recent research, students are lucky if they get one afternoon session in three years, even in ostensibly progressive new universities. Feminist scholarship is not being woven into the main literature of urban academia (cf. Bodman, 1991) through citation, cross-referencing and recommended reading to the extent that one would have expected by now. Indeed, it seemed, from my research, that some senior male planners, both in education and practice, had never even heard of women's issues, or other minority demands – it was as if they were living in a separate world. But I knew that these issues were 'real' and that, in contrast, they had been accepted by other majority and minority organisations, particularly outside the world of construction and the built environment.

Bottom-up change agents

Far from altering the culture of the built environment professions, one result of increased access by women and other minorities (but increased social exclusion too) has been the development of a series of new satellites circling 'Planet Construction' with their own subcultures and organisational structures (see Figure 15.1).

Significant groups include the Centre for Accessible Environments (a disability group, mentioned in chapter 6); Planning Aid for London; London Women and the Manual Trades (LWMT, 1996); the Society of Black Architects (mentioned in chapter 8); Age Concern (mentioned in chapter 5); and the Women's Design Service (WDS, 1997). These groups may be seen as potential prime movers for change, albeit effectively outside the mainstream (cf. Kanter, 1983, p. 296) but containing minority construction professionals.

Many minority professionals who have entered the mainstream professions have not progressed as they might have wished, and may feel marginalised into low-status areas of work. New satellite groupings have developed in response to a sense of being unwelcome and undervalued. Many of the people in such influential satellite groups are strongly socially motivated, but they are not all necessarily town planners. Rather, they are likely to include other built environment professionals, like engineers; pressure group representatives, who are very often of the minority group themselves, such as disabled people; and/or community group residents.

The number of those actively involved is admittedly quite small but they may be seen as a force for change in that they form part of a powerful network of alternative groups and they are often highly productive in publication, research and campaigning. Also, the organisations they run offer a model of alternative management structures, which are often based on a more co-operative, inclusive attitude towards employees at all levels and a greater level of the communication with 'society', particularly when their 'client' is the community, or a disadvantaged or under-represented minority group. Whilst it may be argued these organisations are not commercial enterprises and are therefore irrelevant as models for the construction industry, their ability to achieve a great deal with limited resources shows they have a respect for the 'cost factor', and for efficiency, economy and flat management structures, which might be emulated. One often finds among women's groups a complete lack of the social division between manual trades and professional levels that is found in the mainstream industry.

There is often an emphasis in such groups upon passing the 'knowledge' down to the next generation, for example through mentoring schemes in SOBA, and through work-shadowing programmes for young women surveyors and construction managers. In those branches of the built environment professions where private practice predominates, Black professionals and contractors have set up alternative business networks (cf. B. Grant *et al.*, 1996; Harrison and Davies, 1995). Black built environment professionals have set up a range of network groups, to raise the 'visibility' of all-Black practices and individual practitioners, to counter further 'assimilation' or 'exclusion' (SOBA, 1997).

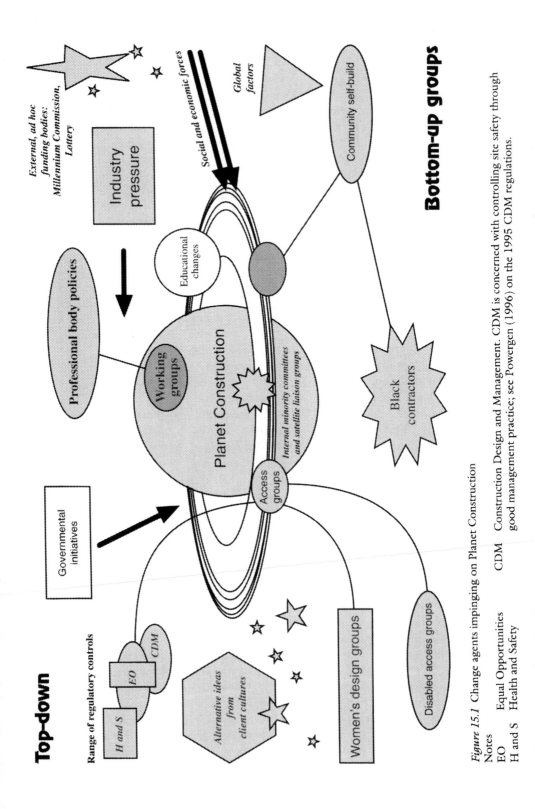

Figure 15.1 Change agents impinging on Planet Construction

Notes
EO   Equal Opportunities
H and S   Health and Safety

CDM   Construction Design and Management. CDM is concerned with controlling site safety through good management practice; see Powergen (1996) on the 1995 CDM regulations.

Creating fusion

Building up a 'critical mass' of minority groups and individuals within the industry is not enough on its own. There is a need for enpowering linkages to be made between external, minority 'bottom-up' groups and top-down 'official' change-agent bodies. Bringing these two bodies together (to continue the nuclear physics analogy) will generate fusion and create a chain reaction that will change the culture of the construction industry. This fusion might occur inside or outside the structures of the built environment professions.

Inside, inter-professional groups such as the Construction Careers Forum, and minority professional groups including CWP (Constructive Women Professionals – significantly, degendered in 1998 and renamed the Equal Opportunities Task Force), together with the sensitive efforts of 'exceptional' contractors are all contributing to change, and act as transmitters and pathways to channel beneficial external influences into the mainstream profession. Bodies and groups representing minority groups within the professions, but strongly linked to 'bottom-up' groups within the community, might act as conduits through which change might flow. But this cannot be done on the cheap without support and resources. Many minority groups are weary of being expected, single-handed, to change the industry with little support from the industry.

Outside and above the world of construction, top down organisations such as, for example, the Lottery, the Millennium Commission, the Sports Council, and the Arts Council exist outside and separate from the construction and built environment world. They have greater powers to require higher accessibility and design standards in buildings than are found under 'normal' legislation (Arts Council, 1996), though, in respect of 'what is built' and 'who' is building it. Thus a new form of social town planning initiative has developed within the world of arts and media funding. Voluntary bodies representing minority groups are likely to be among the beneficiaries of their grants, thus enabling them to produce exemplar schemes in terms of both design and employment practice. This creates fusion between top-down and bottom-up groups, which should, in theory, generate change.

However, some 'women and planning' groups are highly suspicious of the ad hoc Lottery approach to funding, which is based upon competition rather than comprehensive long-term policy, and which may yet prove temporary. Black groups may still be excluded even when other community groups are getting recognition, particularly when it comes to architectural competitions for new schemes. Many are critical of the perceived racism of some housing associations who put Black women professionals on their management committee 'because it looks good', but never actually use Black professionals in construction projects (Harrison and Davies, 1995).

Local government also still has a role to play, in spite of the decline in DLOs (direct labour organisations), which used to be popular employers for women and other minorities, providing support, training, and often the first opportunity for 'real' employment in construction. Nowadays Local Labour in Construction

Initiatives, such as GLIiC (Greenwich) provide employment and information (LWMT, 1996), in association with major contractors and developments in the area, whilst providing liaison with training schemes and local small business initiatives. Getting people involved in construction may prove to be a route into getting them interested in town planning and wider urban policy – indeed NVQs (National Vocational Qualifications) and GNVQs at pre-degree level do not make such a distinction (and do not 'talk down' to so-called underachievers) but cover 'everything', putting detailed construction modules alongside more conceptual urban policy units. Such programmes are crucial in bringing the possibility of built environment education and qualifications right into the community setting.

Many London borough town planning departments make it a requirement of any major planning permission that under a Section 106 Agreement certain equal opportunities measures and design features should be integrated in the development, although such measures have been subject to legal challenge, as *ultra vires*. Likewise, a system of 'zoning bonusing' in some North American states operates in a similar manner. Provision under such agreements might include as part of the development crèche and childcare provision, improved public facilities, improved access and environmental improvements. Clearly, as a process concerned with property, land and development rights, town planning is open to negotiation, because it is highly political and financially driven (Cullingworth and Nadin, 1996). It is unfortunate that often the only way to achieve high social planning goals is through such questionable dealing (little has changed since this was discussed in Greed, 1994).

Creating actual cultural change is difficult. New ad hoc 'social planning' initiatives such as the Single Regeneration Budget programmes and housing association schemes (WDS, 1997), have created new possibilities for the involvement of local people and professionals on specific building projects, with inevitable cultural clashes and re-education on both sides. A range of controversial 'top-down' urban renewal schemes in inner London have created tremendous levels of community involvement and 'bottom up' response, potentially creating 'fusion'. Such schemes include the King's Cross development site, and the 'Guinness Site', Gargoyle Wharf in Wandsworth, where demands have been made for the use of local labour, designers and planning ideas. All this may be a major headache for those managing the development process but it is a sign of the desire of ordinary people to be more involved in built environment and construction matters. Who knows, as a result, some of them or their children (or mothers) may want to enter the built environment professions as a career. Such schemes also provide a setting in which planners and planned may become more aware of each other's culture and priorities, as part of an ongoing learning process.

Organisations such as Planning Aid provide individuals and communities with assistance 'against' the planners, and in putting forward positive alternatives to planning proposals (Parkes, 1996). However, some have not given up on the existing system, and have sought to develop new forms of public participation and involvement of the general public in the planning process, breaking down the 'them and us' syndrome. Traditionally, members of the public have been invited to comment upon the proposals in a new development plan before its adoption, and their views will be sought through public meetings, exhibitions and questionnaires. The process has been much criticised, as it may only attract the more articulate. There may be no real choice of policy on offer. Much depends on what questions are asked and what issues are seen to be relevant to the planning process. Some local authorities have pioneered more meaningful, long-term programmes of gradually working with minority groups in the community, in schools and with voluntary organisations, 'teaching' the participants about how the planning system works and discussing policy alternatives with them, until after several months they are in a better position to participate constructively in the planning process, in contrast to the 'hit and run' approach of one-off participation exercises.

In their new post-Thatcher, more entrepreneurial roles, which may involve them, for instance, in developing partnerships with business, local authority town planners may encounter more 'feminised' cultures (because of the composition of their workforces) in their meetings with retail organisations, banks, and service providers (such as health authorities), all of which may possess alternative business and professional cultures. One of the obstacles to cultural change is fear and an unwillingness among some managers and professionals to admit the need for drastic change. One manager in a large construction organisation confided that he felt very threatened by women coming in, confessing that he got really worried when he realised that equal opportunities did not just affect a few women, but that it might affect him personally, in that he would have to change not only his personal way of life, and 'the whole industry' but even the way he spoke on site to workers and colleagues. Clearly, the penny is dropping at last that mainstream cultural change is the logical result of effective equal opportunities policy. This incident was a refreshing change from speaking to other male professionals who, as female colleagues commented, 'simply don't realise what it's all about' (a key theme) but who could nevertheless speak authoritatively and enthusiatically about equal opportunities when necessary.

Professional education can also provide a site for cultural change. It may be that economic factors (especially reductions in home student grants and increased tuition fees), rather than social concerns about equal opportunities or discrimination, will lead to culture change, as colleges realise that to attract more ethnic minority students they need to meet the needs of minority groups, rather than simply expecting them to 'fit in and put up with it all'. More positively, it may be argued that a wider range of types of students would lead to greater cultural diversity and greater critical mass which might, in turn, attract more minority candidates, and change the culture of construction for the better. For example, as stated (Baty, 1997) over half of students in some central London new universities are ethnic minority.

New cultural combinations and influences are thus created which transform traditional historical and sociological locational associations. For example a significant minority of Middle Eastern students may choose erstwhile depressed British industrial towns, now regenerated by Asian business enterprise, over more prestigious alternatives because 'Bradford is such a wonderful European city, it has everything a Muslim could want'. This city is particularly attractive to would-be built environment professionals, providing both English and foundational built environment education in local colleges. To assume that providing education for ethnic minority students is a means of providing for the deprived in the inner city or a form of charity intended to 'help' backward countries, reflects a paternalistic mentality reminiscent of the heyday of the British Empire. In reality, some overseas and home ethnic minority students are from relatively wealthy professional families. Indeed, students' parents may consider the built environment professions to be an unsuitable, low-status career choice for their sons or daughters, preferring them to go into law, accountancy or medicine. Ironically, whilst the promotion of courses may be done in the name of 'equal opportunities' the hard truth in many a new university is that overseas students pay more than home students. Some construction courses for which applications have fallen are virtually dependent upon getting overseas students to ensure their continued running and thus college jobs and resources.

At the international level, lessons might be learnt on how to change the 'construction culture' in Britain to attract more ethnic minorities, by observing what is happening in relatively more 'neutral' multinational working situations abroad where many cultures meet (cf. Hammuda and Dulaimi, 1997, pp. 27–9) and where Muslim (Middle Eastern), Confucian (Far Eastern) and Judaeo-Christian (Western) styles of management and labour relations come together. More locally, increased harmonisation of labour practices and conditions within the European Union is also bound to result in new perspectives and approaches to site relations, which may benefit minority groups.

Envoi

In this chapter, a diverse range of possibilities and means of change have been presented. It must be admitted that it has proved impossible to answer the question 'How might cultural change be achieved within the planning profession?' But it is hoped that the book has provided a set of examples and illustrations of different ways of going about social town planning. In conclusion, I would argue that one of the most promising ways to break down the physical/social dualism is by the injection into town planning of a wider range of types of people, from a diversity of academic, community, and professional realms, who, ideally, are innocent of how planning 'ought to be'. Combined with this, public pressure, the power of community and minority campaign groups and the influence of education and the media are all means of changing the spirit of planning.

Bibliography

Access Officers' Association (1996) *Briefing Newsletter*, London: Access Officers' Association.

ACE (Access Committee for England) (1997) *Access to Government*, London: ACE.

Age Concern (1993) *Housing for All*, London: Age Concern England.

AGSD (Advisory Group on Sustainable Development) (1997) *Report to The Secretary of State for Scotland*, Edinburgh: HMSO.

Alexander, C. (1965) 'A city is not a tree', *Architectural Forum*, vol. 122, no. 1, April 1965; reprinted in R.T. Le Gates and F. Stout (eds) (1996) *The City Reader*, London: Routledge, part 2, pp. 118–31.

Amin, A and Hausner, J. (1997) 'Interactive governance and social complexity', unpublished report, Perth, Australia (contact J. Hillier, University of Curtin, Perth).

Amos, V. and Parmar, P. (1997) 'Challenging imperial feminism' in H.S. Mirza (ed.) *Black British Feminism*, London: Routledge.

Anderson, S. (1968) (ed.) *Planning for Diversity and Choice*, Cambridge, Mass.: MIT.

Archer, B. and Cooper, C. (1994) 'The positive and negative aspects of tourism', in W. Theobold (ed.) *Global Tourism: The Next Decade*, Oxford: Butterworth–Heinemann.

Argyll and Bute District Council (1985) *Islay, Jura, Colonsay Local Plan* (adopted 20 November 1985), Lochgilphead: Argyll and Bute District Council.

Arnstein, S.R. (1969) 'A ladder of citizen participation', *Journal of the American Institute of Planners*, vol. 35, July, pp. 216–24.

Arts Council (1996) *Equal Opportunities: Additional Guide*, London: National Lottery.

ARU (Les Annales de la Recherche Urbaine) (1997) 'Emplois du Temps', special issue on 'Time Planning', no. 77, December, Paris: Plan Urbain, Ministère de l'équipement du Logement, des Transports et du Tourisme.

Bagilhole, B. (1993) 'How to keep a good woman down: an investigation of institutional factors in the process of discrimination against women academics', *British Journal of Sociology*, vol. 14, no. 3, pp. 262–74.

Bagihole, B. (1997) *Equal Opportunities and Social Policy*, London: Longman.

Bagilhole, B., Dainty, A. and Neale, R. (1996) 'Women in construction: a view of contemporary initiatives in the United Kingdom, and a proposal for international collaborative research', paper presented at the conference 'Gender, Science and Technology', Ahmedabad, India contact Dr Ruth Carter, Open University, Milton Keynes, UK.

Banister, D. (1997) 'Reducing the need for travel', *Environment and Planning B*, vol. 24, no. 3, pp. 437–50.

Barlow Commission (1940) *Royal Commission on the Distribution of the Industrial Population* (Barlow Report), Cmnd 6153, London: HMSO.

Barnes, C. (1991) *Disabled People in Britain and Discrimination*, London: Hurst.

Barton, H. (1998a) 'Design for movement' in C. Greed and M. Roberts (eds) *Introducing Urban Design: Interventions and Responses*, Harlow: Addison, Wesley, Longman, pp. 133–52.

Barton, H. (1998b) 'Planning for access', in C. Greed and M. Roberts (eds) *Introducing Urban Design: Interventions and Responses*, Harlow: Addison, Wesley, Longman.

Barton, H., Guise, R. and Davis, G. (1995) *Sustainable Settlements: A Guide for Planners, Designers and Developers*, Luton: Local Government Management Board in association with University of the West of England, Bristol.

Baty, P. (1997) 'Is the Square Mile colour blind?', *Times Higher Education Supplement*, 7 November, p. 6.

Baxter, J.M. and Usher, M.B. (eds) (1994) *The Islands of Scotland: A Living Marine Heritage*, London: HMSO.

BBC (British Broadcasting Corporation) (1996) *From the Edge: Disability Discrimination Act, 1995*, London: British Broadcasting Corporation Publications Division.

BCODP (British Council of Organisations of Disabled People) (1997), press release in response to a speech made by Andrew Smith, Labour Party Conference, 8 October.

Beall, J. (1997) *A City for All: Valuing Difference and Working With Diversity*, London: Zed Books.

Belfast Healthy City Project (1996) *Belfast City Health Plan Strategy Document*, Belfast: Belfast City Council.

Belloni, C. (1994) 'A woman-friendly city: politics concerning the organisation of time in Italian cities', *Women in the City: Housing Services and Urban Environment*, Paris: Organisation for Economic Co-operation and Development.

Belloni, C. (1996) 'Policies concerning the organisation of urban time in Italy', University of Turin, Italy, Valley d'Aosta Project, summary paper presented, 'Politiche del Tempo in una Prospettiva Europea: Conferenza Internazionale', Eurofem conference, Aosta, Turin, September 1996.

Bhride, K. ni (1987) 'Women and planning: an analysis of women's mobility and accessibility to facilities', unpublished Regional and Urban Planning MA thesis, no. 182, Dublin: University of Dublin.

Bianchini, F. (1995) 'The 24-hour city', *Demos Quarterly*, issue 5, p. 47.

Blowers, A. (ed.) (1993) *Planning for a Sustainable Environment*, London: Town and Country Planning Association and Earthscan.

Bodman, A. (1991) 'Weavers of influence: the structure of contemporary geographic research', *Transaction of the Institute of British Geographers*, vol. 16, no. 1, pp. 21–37.

Booth, C., Darke, J. and Yeandle, S. (eds) (1996) *Changing Places: Women's Lives in the City*, London: Paul Chapman.

Brand, J. (1996) 'Sustainable development: the international, national and local context for women', *Built Environment*, vol. 22, no. 1, pp. 58–71.

Broady, M. (1968) *Planning for People*, London: Bedford Square Press.

Brownill, S. and Halford, S. (1990) 'Understanding women's involvement in local politics: how useful is the formal/informal dichotomy', *Political Geography Quarterly*, vol. 9, no. 4, 396–414.

Bruton, M.J. (1974) (ed.) *The Spirit and Purpose of Planning*, London: Hutchinson (2nd edition, 1984).

BSI (1995) *Sanitary Installations Part I: Code of Practice for Scale of Provision, Selection and Installation of Sanitary Appliances*, no. 6465, London: British Standards Institute (HMSO).

Bucher, R. and Strauss, A. (1961) 'Professions in process', *American Journal of Sociology*, vol. 66, no. 4, pp. 325–34.

Butler, F. (1993) 'The EC's Common Agricultural Policy (CAP)', in J. Lodge (ed.) *The European Community and the Challenge of the Future*, London: Pinter.

Bytheway, B. and Johnson, J. (1990) 'On defining ageism', *Critical Social Policy*, vol. 10, no. 2, issue 29, pp. 27–39.

CADISPA (Conservation and Development in Sparsely Populated Areas)(1995a) *A Blueprint for Rural Europe All About CADISPA*, WWF, Gland, Switzerland.

CADISPA (1995b) *Scotland: The Birth of CADISPA*, WWF, Gland, Switzerland.

CAE (Centre for Accessible Environments) (1998) *Keeping up with the Past: Making Historic Buildings Accessible to Everyone* (video), London: Centre for Accessible Environments.

Callon, M. (1986) 'Some elements of a sociology of translation', in J. Law (ed.) *Power, Action, Belief: A New Sociology of Knowledge?* , London: Routledge, pp. 196–233.

Callon, M. Law, J. and Rip, A. (1986) *Mapping the Dynamics of Science and Technology*, London: Macmillan.

Capital City Partnership (1996a) *Closing the Gap: Creating Sustainable Regeneration in Edinburgh*, City of Edinburgh Council, Strategic Policy, Edinburgh.

Capital City Partnership (1996b) *Making a Difference: The North Edinburgh Strategy*, City of Edinburgh Council, Strategic Policy Edinburgh.

Cavanagh, J. (1998) 'Others and structures in the post(-)rural: degrees of movement and separation', unpublished Ph.D. thesis, University of Exeter.

CEC (Commission of the European Communities)(1985) *Completing the Internal Market*, White Paper, June, Luxembourg: OOPEC

CEC (1990) *Green Paper on the Urban Environment*, COM(90) 218 CEC, Luxembourg: OOPEC.

CEC (1991) *Europe 2000: Outlook for the Development of the Community's Territory*, Directorate-General for Regional Policy and Cohesion, Brussels, COM(91) 452 CEC, Luxembourg: OOPEC.

CEC (1992a) *Proposal for a Directive on the Mobility and Safe Transport to Work of Workers with Reduced Mobility*, Official Journal C15, 21 January 1992, Luxembourg: OOPEC.

CEC (1992b) Treaty on European Union (Maastricht Treaty), Luxembourg: OOPEC.

CEC (1992c) *Towards Sustainability: A European Community Programme of Policy and Action in Relation to the Environment and Sustainable Development*, Fifth Environmental Action Programme, COM(92) 23, Luxembourg: OOPEC.

CEC (1992d) *The Future Development of the Common Transport Policy: A Global Framework for Sustainable Mobility*, COM(92) 494, Luxembourg: OOPEC.

CEC (1992e) *Green Paper on the Impact of Transport on the Environment*, COM(92) 46 final, Brussels, 20 February, Luxembourg: OOPEC.

CEC (1993a) *Report on the Accessibility of Transport to Persons with Reduced Mobility*, COM (93) 433, Luxembourg: OOPEC.

CEC (1993b) *Growth, Competitiveness, Employment: The Challenges and Ways Forward Into the Twenty-First Century* (Delors White Paper), COM(93) 700 final, Brussels, 5 December, Luxembourg: OOPEC.

CEC (1994a) *Community Initiative Concerning Urban Areas*, (URBAN) COM(94) 61 final, Brussels 2 March, Luxembourg: OOPEC.

CEC (1994b) *State of Europe's Environment*, Luxembourg: OOPEC.

CEC (1994c) *Europe 2000+: Cooperation for European Territorial Development*, Directorate-General for Regional Policy and Cohesion, Luxembourg: OOPEC.

CEC (1995a) *Communication from the Commission to the Council, the European Parliament, the Economic and Social Committee and The Committee of the Regions: The Common Transport Policy Action Programme 1995–2000*, COM(95) 302 final, Brussels, 12 July, Luxembourg: OOPEC.

CEC (1995b) *Towards Fair and Efficient Pricing in Transport: Policy Options for Internalising the External Costs of Transport in the European Union*, COM(95) 691, Luxembourg: OOPEC.

CEC (1995c) *Structural Funds Innovatory Measures 1995–1999: Guidelines for the Second Series of Actions under Article 10 of the ERDF Regulation*, doc. XVI/261/95, Luxembourg: OOPEC.

CEC (1995d) *Cohesion Policy and the Environment: Communication from the Commission to the Council, the European Parliament, the Economic and Social Committee and the Committee of the Regions.* COM(95) 509 final, Brussels, 22 November, Luxembourg: OOPEC.

CEC (1995e) *Report from the Commission on the State of Implementation of Ambient Air Quality Directives*, COM(95) 372 final, Brussels, 26 July, Luxembourg: OOPEC.

CEC (1995f) *Amended Proposal for a Council Directive on Ambient Air Quality Assessment and Management*, presented by the Commission pursuant to Article 189a (2) of the EC Treaty, COM(95) 312 final, Brussels, 6 July, Luxembourg: OOPEC.

CEC (1996a) *Social Europe-2/95 Social Dialogue: The Situation in the Community in 1995*, Directorate-General for Employment, Industrial Relations and Social Affairs, Luxembourg: OOPEC.

CEC (1996b) *Progress Report from the Commission on the Implementation of the European Community Programme of Policy and Action in relation to the Environment and Sustainable Development 'Towards Sustainability'*, COM(95) 624 final, Brussels, 10 January, Luxembourg: OOPEC.

CEC (1996c) *European Sustainable Cities: Report by the Expert Group on the Urban Environment*, Luxembourg: OOPEC.

CEC (1996d) *The Citizens' Network: Fulfilling the Potential of Public Passenger Transport in Europe*, Green Paper, Luxembourg: OOPEC.

CEC (1997a) *Social Europe Progress Report on the Implementation of the Medium-Term Social Action Programme 1995–7*, Directorate-General for Employment, Industrial Relations and Social Affairs, Luxembourg: OOPEC.

CEC (1997b) *Agenda 2000 For a Stronger and Wider Union*, Bulletin of the European Union, May, Luxembourg: OOPEC.

CEC (1997c) *Towards an Urban Agenda in the European Union*, Communication COM (97) 197 final, Luxembourg: OOPEC.

CEC (1998a) *Communication on the Social Action Programme 1998–2000*, COM (98) 259, Luxembourg: OOPEC.

CEC (1998b) *Sustainable Urban Development in the European Union: A Framework for Action.* COM (98) 605 final, Luxembourg: OOPEC.

CEC (n.d.) *Europe's Cities: Community Measures in Urban Areas*, Directorate-General for Regional Policy and Cohesion, Luxembourg: OOPEC.

Cecchini, P. (1988) *The European Challenge 1992: The Benefits of a Single Market*, Aldershot: Wildwood House.

Chadwick, G. (1966) 'A systems view of planning,' *Journal of the Town Planning Institute*, vol. 52, no. 5, pp. 184–6.

Cherry, G. F. (1974) *The Evolution of British Town Planning*, London: Leonard Hill.

Churchill, H. and Everitt, A. (1996) *Home from Home: Conversations with Older People in*

Sheltered Housing, University of Northumbria at Newcastle, Social Welfare Research Unit.

CIB (Construction Industry Board) (1994) *Constructing the Team* (The Latham Report), London: Department of the Environment and CIB draft report.

CIB (1996) *Tomorrow's Team: Women and Men in Construction*, Report of Working Group 8 of Latham Committee, London: Department of the Environment and CIB.

CISC (Construction Industry Standing Conference) (1994) *Occupational Standards for Professional, Managerial, and Technical Occupations in Planning, Construction, Property and Related Engineering Services*, London: CISC.

City of London Access Group (1997/8) Minutes of meetings, London: City of London Access Group.

City of Edinburgh Council (1996) *Moving Forward: Strategy*, Edinburgh: Edinburgh City Council.

City of Edinburgh Council (1998) *Environment Strategy: A Quality Environment for a Quality of Life*, Edinburgh: Edinburgh City Council.

City of Liverpool (1996) *Liverpool City Health Plan: Liverpool Healthy City 2000*, Liverpool: Liverpool City Council.

City of Vancouver (1995) *City Plan: Directions for Change*, Vancouver: Vancouver City Council.

City of Westminster (1997) *Draft Corporate Disability Policy*, London: City of Westminster.

Clare, J. (1994) *Living History: The Industrial Revolution*, New York: Random House.

Clarke, R. 1995, 'Fact, not emotion', letter to the editor, *Wanneroo Times* (Australia), 19 December, p. 20

Cloke, P. and Little, J. (eds) (1997) *Contested Countryside Cultures: Otherness, Marginalisation and Rurality*, London: Routledge.

Code, L. (1995) *Rhetorical Spaces*, New York: Routledge.

Coleman, A. (1985) *Utopia on Trial: Vision and Reality in Planned Housing*, London, Hilary Shipman.

Coleman, R. (1992) 'Designing for our future selves', *Access by Design*, no. 58, pp. 10–11

COSLA (Convention of Scottish Local Authorities) (1997) *Scotland's 21 Steps to Sustainability*, Edinburgh: COSLA.

Courtney-Clarke, M. (1986) *Ndebele: The Art of an African Tribe*, London: Rizzoli International Publications.

Courtney-Clarke, M. (1990) *African Canvas*, London: Rizzoli International Publications.

CRE (Commission for Racial Equality) (1991) *Measure of Equality*, Birmingham: Commission for Racial Equality.

Crow, L. (1992) 'Renewing the social model of disability', *Coalition*, July, pp. 5–9.

CSD (Committee on Spatial Development) (1994) 'Principles for a European Spatial Development Policy,' paper to the Informal Council of Spatial Planning Ministers, meeting at Leipzig 21–2 September.

CSD (1997) *European Spatial Development Perspective*, complete draft, Committee on Spatial Development, London: HMSO.

CVS (Community Volunteer Service) (1995) 'Environment', *Scotland's 21 Today*, newsletter, Issue 1.

Cullingworth, J.B. and Nadin, V. (1997) *Town and Country Planning in the UK* (12th edition, first published 1964) London: Routledge.

Cunningham, I. (1994) 'Our irreplaceable heritage', *The Western Review* (Australia), November, pp. 7–8.

Daunt, P. (1991) *Meeting Disability: A European Response*, London: Cassel.

Davidoff, L., L'Esperance, J. and Newby, H. (1976) 'Landscapes with Figures', in A. Oakley and J. Mitchell (eds) *The Rights and Wrongs of Women*, Penguin: Harmondsworth.

Davidoff, P. (1965) 'Advocacy and pluralism in planning', *Journal of the American Institute of Planners*, vol. 31 November; reprinted in A. Faludi (1973) *A Reader in Planning Theory*, Oxford: Pergamon, pp. 277–96.

Davidson, R. and Maitland, R. (1997) *Tourism Destinations*, London: Hodder and Stoughton.

Davies, J.G. (1972) The Evangelistic Bureaucrat: A Study of a Planning Exercise in Newcastle-upon-Tyne, London: Tavistock.

Davies, J.K. and Kelly, M.P. (1993) *Healthy Cities: Research and Practice*, Routledge: London.

Davies, L. (1992) 'Aspects of equality', *The Planner*, vol. 79, no. 3, pp. 14–16.

Davies, L. (1996) 'Equality and planning: gender and disability', in C. Greed (ed.), *Implementing Town Planning: The Role of Town Planning in the Development Process*, Harlow, Essex: Longman.

Davies, L. (1999) *Disability and Planning: An Inclusive Approach* Faculty of the Built Environment occasional paper, Bristol: University of the West of England.

de Boer, E. (1986) *Transport Sociology*, Oxford: Pergamon.

de Graft-Johnson, Ann (1991) 'Where are they? Black women: architecture and the built environment', paper presented to the conference on Gender and Design, London, and published in *Issue: The Magazine of the Design Museum*, Autumn, 1991 (see p. 4).

Deakins, N. and Edwards, J. (1993) *The Enterprise Culture and the Inner City*, London: Routledge.

Dennis, N. (1970) *People and Planning: The Sociology of Housing in Sunderland*, London: Faber & Faber.

Dennis, N. (1972) *Public Participation and Planners' Blight*, London: Faber & Faber.

Department of Health (1998) *Our Healthier Nation: A Contract for Health*, London: HMSO.

DETR (1997a) and annual, *Transport Statistics*, London: HMSO.

DETR (1997b) *British Government Panel on Sustainable Development, Third Report*, London: HMSO.

DETR (1998a) '*Better access planned for new homes*', press release by Construction Minister Nick Raynsford, 3 March, London: HMSO.

DETR (1998b) *Impact of the EU on the UK Planning System*, London: HMSO.

DETR (1998c) *Opportunities for Change: Consultation Paper on a Revised UK Strategy for Sustainable Development*, London: HMSO.

DETR (1998d) White Paper on Transport, London: HMSO.

DoE (Department of the Environment) (1969) *People and Planning: Report of the Committee on Public Participation in Planning* (Skeffington Report), London: HMSO.

DoE (1985) *Access for the Disabled*, Development Control Policy Note 16, London: HMSO.

DoE (1990) *This Common Inheritance: Britain's Environmental Strategy*, Cmnd 1200, London: HMSO.

DoE (1993) Environmental Appraisal of Development Plans: A Good Practice Guide, London: HMSO.

DoE (1994a) Planning Policy Guidance Note 13: Transport, London: HMSO.

DoE (1994b) *Sustainable Development: The UK Strategy*, Cmnd 2426, London: HMSO.

DoE (1994c) *Partnerships in Practice*: Report on 'Partnerships for Change', international conference held in Manchester, September 1993, London: HMSO.

DoE (1995a) *Projections of Households in England to 2016*, London: HMSO.

DoE (1995b) *The Building Regulations 1991: Access and Facilities for Disabled People: New Dwellings*, Approved Document, London: HMSO.

DoE (1995c) *Rural England: A Nation Committed to a Living Countryside*, White Paper, London: HMSO.

DoE (1996) *Indicators of Sustainable Development for the UK*, London: HMSO.

DoE (1997a) *Planning Policy Guidance Note 1 (Revised): General Policy and Principles*, London: HMSO (Previously 1992).

DoE (1997b) *Planning and the Historic Environment*, Planning Policy Guidance Note 15, London: HMSO.

DoE (1997c) *This Common Inheritance: UK Annual Report 1997*, Cmnd 3556, London: HMSO.

Domingo, P.B. and Lopez, C.M. (eds) (1995) *Del Patio a la Plaza: Las Mujeres en Las Sociedades Mediterraneas* (From patio to town square, women in Mediterranean societies), Granada: Universidad de Granada, Spain.

Donnison, D. 1975 'The age of innocence is past: some ideas about urban research and planning', *Urban Studies*, vol. 12, pp. 263–72.

DoT (Department of Transport) (1994) *Low-Floor Bus Survey* London: HMSO.

Dredge, D. and Moore, S. (1992) 'A methodology for the integration of tourism in town planning' *Journal of Tourism Studies*, vol. 3, no. 1, pp. 8–21.

Druker, J., White, G., Hegewisch, A., and Mayne, L. (1996) 'Between hard and soft HRM: human resource management in the construction industry', *Construction Management and Economics*, vol. 14, pp. 405–16.

EEA (European Environment Agency) (1995) *Europe's Environment: The Dobris Asssessment*, Luxembourg: OOPEC.

EHTF (English Historic Towns Forum) (1994) *Getting It Right: A Guide to Visitor Management in Historic Towns*, London: EHTF.

EIS (European Information Service) (1998) 'Social policy until the year 2000', *European Information Service*, no. 190, pp. 46–9, London: Local Government International Bureau, European Information Service.

ESCTC (European Sustainable Cities and Towns Campaign) (1994) *Charter of European Cities and Towns Towards Sustainability* (the Aalborg Charter), Brussels

Eurofem (1998) *Gender and Human Settlements: Local and Regional Sustainable Human Development from a Gender Perspective*. Proceedings of the Eurofem International Conference held at Hämeenlinna, Finland, June 1998, ed. Liisa Horelli. Helsinki: Yliopistopaino, Pikapaino.

European Institute of Urban Affairs (1992) *Urbanisation and the Functions of Cities in the European Community*, Regional Development Studies, no. 4, CEC Directorate-General for Regional Policies. Prepared by the European Institute of Urban Affairs, Liverpool John Moores University.

Fagan, G. R. (1997a) *The CADISPA Project: Final Activity Report 1997* Glasgow: University of Strathclyde and CADISPA.

Fagan, G. R. (1997b) *New Ideas in Rural Development, no. 3: Involving Rural Communities : The CADISPA Approach*, Edinburgh: The Scottish Office Central Research Unit.

Fainstein, N.I. and Fainstein, S.S. (1979) 'New debates in urban planning: the impact of Marxist theory within the United States', *International Journal of Urban and Regional Research*, vol. 3, no. 3 September. Reprinted in Paris (1982), pp. 147–73.

Faludi, A (1972) *The Specialist Versus Generalist Conflict*, Blackwell: Oxford.

Faludi, A. (1973) (ed.) *A Reader in Planning Theory*, Oxford: Pergamon.

Finkelstein, V. (1993) 'Disability: a social challenge or an administrative responsibility?' in J. Swain, V. Finkelstein, S. French and M. Oliver (eds) *Disabling Barriers: Enabling Environments*, London: Sage Publications in association with the Open University.

Fleck, J. (1992) 'Access for people with disability', unpublished conference paper, London: City of London planning office.

Fleck, J. (1996) letter, *Planning Week*, 25 April, p. 12.

FoE (Friends of the Earth) (n.d.) 'Sustainable Europe: A report from the UK', unpublished report, London: FoE.

Fogelsong, R.E. (1996) 'Planning the capitalist city', in S. Campbell and S. Fainstein (eds) *Readings in Planning Theory*, Oxford: Blackwell.

Ford, J. and Sinclair, R. (1987) *Sixty Years On: Women Talk About Old Age*, London: Women's Press.

Forester, J. (1993a) *Critical Theory: Public Policy and Planning Practice*, Albany, New York: State University of New York Press.

Forester, J. (1993b), 'Beyond dialogue to transformative learning: how deliberative rituals encourage political judgement in community participation processes', working paper, Haifa: Center for Urban and Regional Studies.

Forester, J. (1996) 'Beyond dialogue to transformative learning', in S. Esquith (ed.) *Democratic Dialogues: Theories and Practices,* Poznan: University of Poznan.

Forward Scotland (1997) *Annual Review of 1997*, Glasgow: Forward Scotland.

Frazer, D. (1997) 'Clearing the land we love', *The Age* (Australia), 14 July, p. A13.

Frechette, L.A., (1996) *Accessible Housing*, New York: McGraw–Hill.

Friends of the Earth Scotland (1995) *Towards Sustainable Europe: The Study*, Edinburgh: FoE.

Friends of the Earth Scotland (1996) *Towards a Sustainable Scotland: A Discussion Paper*, Edinburgh: FoE.

Fudge C. (1995) *International Healthy and Ecological Cities Congress: Our City, Our Future, Rapporteur's Report*, Copenhagen: WHO.

Fudge C. and Rowe J. (1997) *Urban Environment and Sustainability: Developing the Agenda for Socio-Economic Environmental Research*. Research Report for DG XII, Bristol: University of the West of England.

Gale, A. (1995) 'Women in construction' in D. Langford, M.R. Hancock, R. Fellows and A. Gale (1994) *Human Resources in the Management of Construction*, Harlow: Longmans, pp. 161–87.

Geddes, P. (1915) *Cities in Evolution: An Introduction to the Town Planning Movement and the Study of Civics*, London: Ernest Benn (reprinted 1968).

George, S. (1991) *Politics and Policy in the European Community*, Oxford: Oxford University Press.

Giddens, A. (1994a) 'Living in a post-traditional society', in U. Beck, A. Giddens and S. Lash *Reflexive Modernisation*, Cambridge: Polity Press, pp. 56–109.

Giddens, A. (1994b) *Beyond Left and Right*, Cambridge: Polity Press.

Gilroy, R. (1993) *Good Practices in Equal Opportunities*, Aldershot: Avebury.

Gilroy, R. (1996a) 'Buiding routes to power', *Local Economy*, vol. 11, no. 3, pp. 248–58.

Gilroy, R. (1996b) Building A Vision of a City for Everyone: Planning Vancouver, Department of Town and Country Planning Working Paper, no. 54, Newcastle: University of Newcastle.

Gilroy, R. and Castle, A. (1995) *Planning For Our Own Tomorrow: Planning For Older People*, Department of Town and Country Planning working paper no. 48, Newcastle: University of Newcastle upon Tyne.

Gilroy, R. and Woods, R. (1994) *Housing Women*, London: Routledge.

Girouard, M. (1990) *The English Town*, New Haven: Yale University Press.

Glasgow Healthy City Project (1995) *Working Together for Glasgow's Health: Glasgow City Health Plan*, Glasgow: Glasgow City Council.

GLC (Greater London Council) (1984) *Planning for the Future of London*, London: GLC

Goodall, B. and Stabler, M. J. (1997) 'Principles influencing the determination of environmental standards for sustainable tourism', in M.J. Stabler (ed.) *Tourism and Sustainability: Principles to Practice*. Wallingford: CAB International.

Goodman, R. (1972) *After the Planners*, Harmondsworth: Penguin.

Goodwin, M. (1998) 'The governance of rural areas: some emerging research issues and agendas', *Journal of Rural Studies* vol. 14, no. 1, pp. 5–12.

Goodwin, P. (1997) 'Road traffic growth and the dynamics of sustainable transport policies', in B. Cartledge (ed.) (1997) *Transport and the Environment*, Oxford: Oxford University Press.

Goodwin, P., Bailey, J. M., Bribourne, R. H., Clarke, M. I., Donnison, J. R., Render, T. E. and Whiteley, G. K. (1983) *Subsidised Public Transport and the Demand for Travel*, Aldershot: Gower.

Goodwin, P., Hallett, S., Kenney, F. and Stokes, G. (ed.) (1991) *Transport: The 'New Realism'*, Report to the lees Jeffreys Road Fund, Transport Studies Unit, Oxford: University of Oxford.

Gore, T. and Nicholson, T. (1991) 'Models of the land development process: a critical review', *Environment and Planning A*, vol. 23, pp. 705–30.

Grant, B., E. Owusu, A. de Graft-Johnson, A. Long, A. English, C. Shokoya and E. Kalu (1996) *Building E=Quality: Minority Ethnic Construction Professionals and Urban Regeneration*, committee report, London: House of Commons.

Grant, M. (1996) Presentation to Tourism Society seminar, 'Working together for visitor management', Cambridge: English Tourist Board.

Grant, M., Human, B. and Le Pelley, B. (1995) 'Tourism and town centre management', *Insights*, Journal of the English Tourist Board, July, pp. A21–A25.

Greed, C. (1988) Is more better?: with reference to the position of women chartered surveyors in Britain, *Women's Studies International Forum*, vol. 11, no. 3, pp. 187–97.

Greed, C. (1991) *Surveying Sisters: Women in a Traditional Male Profession*, London: Routledge.

Greed, C. (1994) *Women and Planning: Creating Gendered Realities*, London: Routledge.

Greed, C (ed.) (1996a) *Investigating Town Planning: Changing Perspectives and Agendas*, Harlow: Longman.

Greed, C. (ed.) (1996b) *Introducing Town Planning*, Harlow: Longmans, 2nd Edition

Greed, C. (1996c) 'Planning for women and other disenabled groups, with reference to the provision of public toilets in Britain', *Environment and Planning A*, vol. 28, pp. 573–88.

Greed, C. (1997a) 'Bad timing means no summer in the cities', *Planning*, 14 March, issue 1209, pp. 18–19.

Greed, C. (1997b) 'Cultural change in construction', *Proceedings of the ARCOM Conference, Cambridge*, 15–17 September 1997, Association of Researchers in Construction Management.

Greed, C. (1999) *The Changing Composition of the Construction Professions: Statistical Appendix*, Faculty of the Built Environment occasional paper, Bristol: University of the West of England.

Greed, C. (forthcoming) *The Changing Composition and Culture of the Construction Professions*.

Greed, C. and Roberts, M. (eds) (1998) *Introducing Urban Design: Interventions and Responses*, Harlow: Addison, Wesley, Longman.

Griffiths, R. (1986) 'Planning in retreat? Town planning and the market in the eighties', *Planning Practice and Research*, no. 1, pp. 3–7.

Grove White, R. (1997) 'The environmental valuation controversy', in J. Foster (ed.) *Valuing Nature? Economics, Ethics and the Environment*, London: Routledge.

Guinchard, C.G. (ed.) (1997) *Swedish Planning: Towards Sustainable Development* Stockholm: The Swedish Society for Town and Country Planning.

Gunn C.A. (1994) 'The emergence of effective tourism planning and development', in A. V. Seaton (ed.) *Tourism: The State of the Art*, Chichester: Wiley.

Guy, S. and Marvin, S. (forthcoming), 'Towards a new paradigm of transport planning?', *Town Planning Review*.

Habermas, J. (1984) *The Theory of Communicative Action*, vol. 1, Boston: Beacon Press.

Habermas, J. (1987) *The Theory of Communicative Action*, vol. 2, Boston: Beacon Press.

Habermas, J. (1992) *Postmetaphysical Thinking*, Cambridge: Polity Press.

Hagerstrand, T. (1968) *Innovation Diffusion as a Spatial Process*, Chicago: University of Chicago Press (translated by R. Allan Pred from the Swedish edition).

Hague, C. (1984) *The Development of Planning Thought*, London: Hutchison.

Hague, C. (1996) 'The development and significance of the RTPI's role in planning education', paper presented at the ACSP/AESOP Joint Congress (available from H. Thomas, University of Wales, Cardiff).

Halford, S. (1989) 'Spatial divisions and women's initiatives in British local government', *Geoforum*, vol. 20, no. 2, pp. 161–74.

Hall, P. (1992) *Urban and Regional Planning*, London: Routledge.

Hall, P., Gracey, H., Drewett, R. and Thomas, D. (1973) *The Containment of Urban England* (2 vols), London: George Allen and Unwin.

Halliday, J. (1997) 'Children's services and care: a rural view', *Geoforum*, vol. 28, no. 1, pp. 103–19.

Hammuda, I. and Dulaimi, M. (1997) 'The effects of the situational variables on the leadership styles in construction projects', in *Proceedings of the ARCOM Conference, Cambridge 15–17 September 1997* pp. 22–31, 27–9.

Haraway, D. (1992) 'The promises of monsters: a regenerative politics for inappropriate/d others', in L. Grossberg, C. Nelson and P. Treichler (eds) *Cultural Studies*, London: Routledge, pp. 295–337.

Harper, S. and Laws, G. (1995) 'Rethinking the geography of ageing', *Progress in Human Geography*, vol. 19, no. 2, pages 199–221.

Harrison, M. (1975) 'British town planning ideology and the Welfare State', *Journal of Social Policy*, vol. 4, no. 3, pp. 259–74.

Harrison, M. and Davies, J. (1995) *Constructing Equality: Housing Associations and Minority Ethnic Contractors*, London: Joseph Rowntree Foundation.

Hartrop, D. (1998) 'Accessible and adaptable housing', undergraduate dissertation, Bristol: University of the West of England, Faculty of the Built Environment.

Harvey, D. (1973) *Social Justice and the City*, London: Edward Arnold.

Harvey D. (1996) 'On planning the ideology of planning', in S. Campbell and S. Fainstein (eds) *Readings in Planning Theory*, Blackwell, Oxford.

Hazel, G. (1997) *Transport for the Twenty-First Century: The Lessons of Edinburgh, Or People, Chairs and Cars in Transport and the Environment*, discussion report, London: Design Council.

HBF (House Builders' Federation) (1995) 'The application of building regulations to help disabled people in new dwellings in England and Wales', unpublished consultation paper, London: HBF

Healey, P. (ed.) (1982) *Planning Theory: Prospects for the 1980s,* Oxford: Pergamon.

Healey, P. (1983) *Local Plans in British Land Use Planning,* Oxford: Pergamon.

Healey, P. (1991) 'Debates in planning thought', in H. Thomas and P. Healey (eds) *Dilemmas of Planning Practice,* Aldershot: Avebury.

Healey, P. (1992), 'Planning through debate: the communicative turn in planning theory and practice', *Town Planning Review,* vol. 63, no. 2, pp. 143–62.

Healey, P (ed.) (1996a) *Negotiating Development* London: Spon.

Healey, P. (1996b) 'Consensus-building across difficult divisions: new approaches to collaborative strategy-making', *Planning Practice and Research,* vol. 11, no. 2, pp. 207–16.

Healey, P (1997) *Collaborative Planning,* Basingstoke: Macmillan.

Healey, P and Hillier, J. (1996), 'Communicative micropolitics: a story of claims and discourses', *International Planning Studies,* vol. 1, no. 2, pp. 165–84.

Healey, P. and Underwood, J. (1978) 'Professional ideals and planning practice', *Progress in Planning,* vol. 2, pp. 73–127

Healy, A. (1997) 'Europe needs you', *Planning,* 10 October, p. 15.

Heeley, J. (1981) 'Planning for tourism in Britain', *Town Planning Review,* vol. 52, pp. 61–79.

Henckel, Dietrich (1996) 'Time in the city: politicalisation of time in German society' Deutches Institut für Urbanistik, Berlin, working paper presented at 'Politiche del Tempo in una Prospettiva Europea: Conferenza Internazionale', Eurofem conference, Aosta, Turin, September 1996.

Higgins, M.D. (1994) 'The economy of the arts: the big picture', speech at the conference 'The Economy of the Arts', Dublin, December 1994.

Hillier, J. (1997) 'The role of procedural justice in determining the effectiveness of citizen involvement in planning', unpublished final report for ARC large grant no. A79602576.

Hillier, J. (1998) 'Imagined value: the poetics and politics of place', in A. Madanipour, A. Hull and P. Healey (eds) *Shaping Places: Conceptions and Directions for Spatial Planning,* London: Routledge.

Hillier, J. and van Looij, T. (1997a), 'Spoken from nowhere? Pragmatic planning and the representation of nature', paper presented at the Environmental Justice conference, University of Melbourne.

Hillier, J. and van Looij, T. (1997b) 'Who speaks for the poor?', *International Planning Studies,* vol. 2, no. 1, pp. 7–25.

Hine, J. (forthcoming) 'Transport policy' in P. Allmendinger and M. Chapman, (eds), *Planning Beyond 2000,* London: John Wiley.

Hogwood, B. (1997), 'The machinery of government 1979–1997', *Political Studies,* vol. 45, no. 4, pp. 704–15.

Holdsworth, A. (1988) *Out of the Doll's House,* London: BBC Publications.

House of Commons (1997a) Treaty of Amsterdam, Cmnd 3780, London: HMSO.

House of Commons (1997b) Scottish Parliament Bill London: HMSO.

House of Commons (1997c) *Scotland's Parliament,* Cmnd 3658, London: HMSO.

House of Lords (1995) *Report from the Select Committee on Sustainable Development,* vol. 1, London: HMSO.

Howard, E. (1898) *Tomorrow: A Peaceful Path to Real Reform,* London: Swan Sonnenschein.

Howitt, R. (1994) *SIA: Sustainability and the Narratives of Resource Regions,* School of Earth Sciences occasional paper, Sydney: Macquarie University.

Hughes, A. (1997) 'Rurality and cultures of womanhood: domestic identities and moral order in village life', in P. Cloke and J. Little (eds) *Contested Countryside Cultures: Otherness, Marginalisation and Rurality*, London: Routledge.

Hull, G.T., Bell Scott, P. and Smith, B. (1982) *All the Women are White, All the Blacks are Men, but Some of Us are Brave*, London: The Feminist Press.

Human, B. (1994) 'Visitor management in the public planning policy context: a case-study of Cambridge', *Journal of Sustainable Tourism*, vol. 2, no. 4, pp. 221–30.

ICLEI (International Council for Local Environmental Initiatives) (1995) *European Local Agenda 21 Planning Guide*, COM, Luxembourg: OOPEC.

Ilbury, B. (1998) 'The challenge of agricultural restructuring in the European Union', in D. Pinder (ed.) *The New Europe*, London: Wiley.

Imrie, R. (1996) *Disability and the City: International Perspectives*, London: Paul Chapman.

Innes, J. (1995) 'Planning theory's emerging paradigm: communicative action and inter-active practice', *Journal of Planning Education and Research*, vol. 14, no. 3, pp. 183–90.

Ismail, A. (1998) 'An investigation of the low representation of black and ethnic minority professionals in contracting', unpublished, special research project, Bristol: University of the West of England.

Jacobs, J. (1964) *The Death and Life of Great American Cities: The Failure of Town Planning*, Harmondsworth: Penguin.

Jenks, M., Burton, E. and Williams, K. (eds) (1996) *The Compact City: A Sustainable Urban Form?*, London: Spon.

Jessop, B. (1995) 'The regulation approach, governance and post-Fordism', *Economy and Society*, vol. 24, no. 3, pp. 307–34.

Jones, O. and Little, J. (forthcoming) 'Rural Challenge(s): partnership and new rural governance', *Journal of Rural Studies*.

John Wheatley Centre (1997) *Working for Sustainability : An Environmental Agenda for A Scottish Parliament, Final Report*, Edinburgh: John Wheatley Centre.

Jones, R. (1982) *Town and Country Chaos: A Critical Analysis of the British Planning System*, London: The Adam Smith Institute.

Kanter, R.M. (1977) *Men and Women of the Corporation*, New York: Basic Books.

Kanter, R.M. (1983) *The Change Masters: Corporate Entrepreneurs at Work, Counterpoint*, London: Unwin.

Keating M (1993) *The Earth Summit Agenda for Change: a Plain Language Version of Agenda 21 and the Other Rio Agreements*, Geneva: Centre for our Common Future.

Keeble, L. (1952) *Principles and Practice of Town and Country Planning*, London: Estates Gazette (2nd edition, 1959).

King, R. (1998) 'From guestworkers to immigrants: labour migration from the Mediterranean periphery', and 'Post-oil crisis, post-Communism: new geographies of international migration', in D. Pinder (ed.) *The New Europe*, London: Wiley.

Kirk, G. (1980) *Urban Planning in a Capitalist Society*, London: Croom Helm.

Kirk-Walker, S. (1997) *Undergraduate Student Survey: A Report of the Survey of First Year Students in Construction Industry Degree Courses*, York: Institute of Advanced Architectural Studies (commissioned by CITB).

Kitchen, T. (1997) *People, Politics, Policies and Plans*, London: Paul Chapman.

Krishnarayan, V. and Thomas, H. (1993) *Ethnic Minorities and the Planning System*, London: RTPI.

Labour Party (1997) *Election Manifesto 1997*, London: The Labour Party.

Landry, C. and Bianchini, F. (1995) *The Creative City*, London: Demos.

Larsen, E. (1958) *Atomic Energy: The Layman's Guide to the Nuclear Age*, London: Pan.

Larson, M.S. (1977) *The Rise of Professionalism*, Berkeley, CA, University of California Press.

Laws, G. (1994) 'Age contested meanings, and the built environment', *Environment and Planning A*, vol. 26, pp. 1787–802.

Leather, P. (1997) 'Providing liveable environments for an increasingly elderly population' paper given to the conference 'How Shall We Live'. Mimeo available from Town and Country Planning Association, 17 Carlton House Terrace, London, SW1Y 5AS.

Lees, A. (1993) *Enquiry into the Planning Systems in North Cornwall District*, London: HMSO.

LGIU (1995) *The Disability Discrimination Act 1995*: A Guide for Local Authorities, London: LGIU (Local Government Information Unit).

LGMB (1993a) *Local Agenda 21 Earth Summit: Rio '92 Agenda 21: A Guide for Local Authorities in the UK*, London: LGMB.

LGMB (Local Government Management Board) (1993b) *A Framework for Local Sustainability*, London: LGMB.

LGMB (1994) *Local Agenda 21: Principles and Process: A Step by Step Guide*, London: LGMB.

LGMB (1996) *Women and Sustainable Development: Local Agenda 21 Round Table Guidance* London: LGMB.

LGMB/RTPI (Local Government Management Board/Royal Town Planning Institute) (1992) *Planning Staffs Survey 1992*, London: LGMB/RTPI.

Lind, E. and Tyler, T. (1988) *The Social Psychology of Procedural Justice*, New York: Plenum Press.

Lindsay, Earl of (1996) *Making Sustainability Our Business*, Edinburgh: Scottish Office

Little, J. (1986) 'Feminist perspectives in rural geography: an introduction', *Journal of Rural Studies*, vol. 2, no. 1, pp. 1–8.

Little, J. (1991) 'Women and the rural labour market', in T. Champion and C. Watkins (eds) *People in the Countryside*, London: Paul Chapman.

Little, J. (1994a) *Gender, Planning and the Policy Process*, Oxford: Elsevier Pergamon.

Little, J. (1994b) 'Women's initiatives in town planning in England: a critical review', *Town Planning Review*, vol. 65, no. 3, pp. 261–76.

Little, J. (1997) 'Employment marginality and women's self-identity', in P. Cloke and J. Little (eds) *Contested Countryside Cultures: Otherness, Marginalisation and Rurality*, London: Routledge.

Little, J. and Austin, P. (1996) 'Women and the rural idyll', *Journal of Rural Studies*, vol. 12, no. 2, pp. 101–11.

Little, J., Clements. J. and Jones, O. (1998) 'Rural Challenge and the changing culture of rural regeneration policy'. in N. Oatley (ed.) *Cities, Economic Competition and Urban Policy*, London: Paul Chapman.

Lock, D. (1998) 'Compassion finds a sustainable slant', *Planning*, 1 May, pp. 16–17.

Long, J. (1994) 'Local authority tourism strategies: a British appraisal', *Journal of Tourism Studies*, vol. 5, no. 2, pp. 17–23.

Long, P. (1994) 'Perspectives on partnership organisations as an approach to local tourism development', in A. V. Seaton (ed.) *Tourism: The State of the Art*, Chichester: John Wiley.

Lonsdale, S. (1990) *Women and Disability: The Experience of Physical Disability Among Women*, London: Macmillan.

Lord Provost's Commission (1998) *The Lord Provost's Commission on Sustainable Development for the City of Edinburgh, Presented to the Lord Provost, July 1998*, Edinburgh: The Lord Provost's Commission.

Lorenz, C. (1990) *Women in Architecture: A Contemporary Perspective*, London: Trefoil.

Lothian Regional Council (1996) *Cycling in Lothian: 2000 Moving Forward*, Edinburgh: Lothian Regional Council.

Love, N. (1995) 'What's left of Marx?', in S. White (ed.) *The Cambridge Companion to Habermas*, Cambridge: Cambridge University Press, pp. 46–66.

LRN (London Regeneration Network) (1997) *Still Knocking at the Door* Report of the Women and Regeneration seminar, London: May 1997, LRN with London Voluntary Service Council.

LWMT (London Women and Manual Trades) (1996) *Building Careers: Training Routes for Women*, London: LWMT

McAllister, G. and McAllister, E. (eds) (1941) *Town and Country Planning* London: Faber & Faber.

McAuslan, J.P. (1980) *The Ideologies of Planning Law*, Oxford: Pergamon.

McCafferty, P. (1994) *Living Independently: A Study of the Housing Needs of Elderly and Disabled People*, Housing Research Summary no. 28, London: HMSO for DoE.

McDowell, L. (ed.) (1997) *Undoing Place? A Geographical Reader*, London: Arnold.

McLaren, D. (1998) *Tomorrow's World: Britain's Share in a Sustainable Future*, London: FoE and Earthscan.

McLaughlan, G. (1997) Address to the NSCA conference by G. McLaughlan, Chair of the Lord Provost's Commission on Sustainable Development, Glasgow, 22 October.

McLoughlin, J.B. (1965) 'The planning profession: new directions', *Journal of the Town Planning Institute*, vol. 51, pp. 258–61.

McLoughlin, J.B. (1969) *Urban and Regional Planning*, London: Faber.

McNamara, K. (1996) 'Working together for visitor management in towns', *Tourism*, Journal of the Tourism Society, vol. 89, p. 17.

McNeill, A.G. (1993) 'Citizen participation in the planning process: a case study of the city of Vancouver's project on ageing', unpublished masters degree thesis, Vancouver: University of British Columbia.

Macpherson, C.B. (1977) *The Life and Times of Liberal Democracy*, Oxford: Oxford University Press.

Maitland, R. (1997) 'Cities, tourism and mixed uses', in A. Coupland (ed.) *Reclaiming the City*, London: Spon.

Manley, S. (1998) 'Creating accessible environments', in C.Greed and M. Roberts (eds), *Introducing Urban Design: Interventions and Responses*, Harlow: Addison, Wesley, Longman, pp. 153–67, and Appendix II, 'Disability'.

Marsden, T., Murdoch, J., Lowe, P., Munton, P. and Flynn, A. (1993) *Constructing the Countryside*, London: UCL Press.

Martin, J., Meltzer, H. and Elliott, D. (1988) *The Prevalence of Disability among Adults*, OPCS Surveys of Disability in Great Britain, Report 1, London: HMSO.

Massey, D. (1984) *Spatial Divisions of Labour: Social Structures and the Geography of Production*, London: Macmillan.

Massey, D. (1991) 'The political place of locality studies', *Environment and Planning A*, vol. 23, pp. 267–82.

Massey, D. (1993) 'Politics and space/time', in M. Keith and S. Pile (eds), *Place and the Politics of Identity*, London: Routledge, pp. 141–61.

Matarasso, F. (1997) *'Use or Ornament?' The Social Impact of Participation in the Arts*, Nottingham and Bournes Green: Comedia.

Matarasso, F. and Halls, S. (eds) (1996) *The Art of Regeneration*, Nottingham and Bournes Green, City of Nottingham and Comedia.

Matrix (1979) 'Women and Space', conference held in London, organised by Matrix, proceedings subsequently formed basis of chapter 7, 'Working with Women' by Frances Bradshaw in Matrix (1984) *Making Space: Women and the Man Made Environment* London: Pluto, pp. 89–105.

Matrix (1980) 'Home Truths' Travelling exhibition organised by Matrix, toured around the country 1980–83 approximately.

Matrix (1984) *Making Space, Women and the Man Made Environment*, London: Pluto.

Matrix (1993) *Gender Issues in the Decision-Making Process With Regard to Urban Space and Housing*, EC funded, co-ordinator Roland Mayerl of 'Habitat et Participation', in collaboration with five other EC-based organisations, Brussels: EU).

Matrix (1994a) 'Gender and cultural issues in urban housing: United Kingdom', paper presented to conference 'Emancipation as Related to Physical Planning and Housing', organised by the Netherlands Institute for Physical Planning and Housing (Emancipation section), held in Briebergen, Netherlands, 1994 (reported in Ottes *et al.*, 1995).

Matrix (1994b) 'Black women in architecture from a UK perspective' paper presented at 'Urbanisation and the Environment', conference held at United Nations Headquarters, Nairobi, organised by the Mazingira Institute, including international seminar on gender aspects. Conference proceedings published Nairobi: Mazingira Institute in association with United Nations.

Matrix (1996) *Women in Architecture*, London: Arts Council.

Matsukawa, J. (1994) 'Tokyo Travel: urban space to move around Tokyo', paper to the OECD conference 'Women in the City: Housing, Services and the Urban Environment', Paris, 4–6 October.

Mazingira Institute (1994) 'Urbanisation and the Environment', conference held at United Nations Headquarters, Nairobi, organised by the Mazingira Institute, including international seminar on gender aspects. Conference proceedings published Nairobi: Mazingira Institute in association with United Nations.

Mazza, L. and Rydin, Y. (eds) (1997) 'Urban sustainability: discourses, networks and policy tools', *Progress in Planning*, vol. 47, pp. 1–74.

Mercer, C. (1996) 'By accident or design? Can culture be planned?' in F. Matarasso and S. Halls (eds) *The Art of Regeneration*, Nottingham and Bournes Green: City of Nottingham and Comedia.

Millerson, G. 1964 *The Qualifying Associations*, London: Routledge & Keegan Paul.

Mirza, H.S. (1992) *Young, Female* and *Black*, London: Routledge.

Mirza, H.S. (ed.) (1997) *Black British Feminism*, London: Routledge

Montgomery, J. (1994) 'The evening economy of cities', *Town and Country Planning*, vol. 63, no. 11, pp. 302–7.

Mongomery, J. and Shornley, A. (eds) (1990) *Radical Planning Initiatives: New Directions in Planning for the 1990s*, Aldershot: Gower.

Morley, L. (1994) 'Glass ceiling or iron cage: women in UK academia', *Gender, Work and Organisation*, vol. 1, no. 4, pp. 194–204.

Morphet, J. (1997) 'Enter the ESDP: plan sans fanfare', *Town and Country Planning*, October, pp. 265–7.

Moxon-Browne, E. (1993) 'Social Policy', in J. Lodge (ed.) *The European Community and the Challenge of the Future*, London: Pinter.

Murdoch, J. and Marsden, T. (1995) 'The spatialisation of politics: local and national actor-spaces in environmental conflict', *Transactions of the Institute of British Geographers*, vol. 20, pp. 368–80.

Murdock, J. (1997) 'Inhuman/nonhuman/human: actor network theory and the prospects of nondualistic and symmetrical perspective on nature and society', *Planning and Environment D*, vol. 15, no. 6, pp. 731–56.

NPF (National Planning Forum) (1998) *Planning for Tourism*. London: LGA.

Newman, O. (1973) *Defensible Space: People and Design in the Violent City*, New York and London: Architectural Press.

Nicholson, D. (1991) 'Planners' skills and planning practice', in H. Thomas and P. Healey (eds) *Dilemmas of Planning Practice*, Aldershot: Avebury.

Norton, A., Stoten, B. and Taylor, H. (1986) *Councils of Care: Planning a Local Government Strategy for Older People*, London: Centre for Policy on Ageing.

Nugent, N. (1991) *The Government and Politics of the European Community*, Basingstoke: Macmillan.

O'Connor, J. (1993) 'Manchester and the millennium: whose culture, whose civilisation?', *Regenerating Cities*, no. 5, pp. 17–18.

OECD (Organisation for Economic Co-operation and Development) (1996) *Tourism Policy and International Tourism in OECD Countries 1993–1994*, Paris: OECD.

Oliver, M. (1996) *Understanding Disability: From Theory to Practice*, London: Macmillan.

ONS (Office of National Statistics; was OPCS) (1996, 1997, 1998 and annual) *Social Trends*, London: HMSO.

OU (Open University) (1998) *New Forms of Partnership*, BBC 2 television programme, part of OU urban studies unit, Milton Keynes: Open University in assocation with British Broadcasting Corporation.

OPCS (Office of Population Censuses and Surveys) (1989) *Surveys of Disability in Great Britain*, London: HMSO.

OPCS (1992) *Labour Force Quarterly Bulletin*, Summer–Autumn 1992, London: HMSO.

Ostner, I. (1993) 'Whose solidarity with whom?', in P. Kaim-Caudle, J. Keithley and A. Mullender (eds) *Aspects of Ageing*, London: Whiting & Birch.

Ottes, L. Poventud, E. Van Schendelen, M. and Segond von Banchet, G. (1995) *Gender and the Built Environment*, Assen 'Emancipation: as Related to Physical Planning and Housing': Van Gorcum. Based on the conference organised by the Netherlands Institute for Physical Planning and Housing (Emancipation section), held in Briebergen, Netherlands, 1994.

Owens, S. (1995) 'From predict and provide to predict and prevent? pricing and planning in transport policy', *Transport Policy*, vol. 2, no. 1, pp. 43–50.

Palfreyman, T. (1998) *Designing for Accessibility: An Introductory Guide*, London: Centre for Accessible Environments.

Paris, C. (ed.) (1982) *Critical Readings in Planning Theory*, Oxford: Pergamon Press.

Parkes, M. (1996) *Good Practice Guide to Community Planning and Development*, London: LPAC (London Planning Aid Consortium).

Parkin, F. (1979) *Marxism and Class Theory: A Bourgeois Critique*, London: Tavistock.

Parry, G., Moyser, G. and Day, N. (1992) *Political Participation and Democracy in Britain*, Cambridge, Cambridge University Press.

PDD (Planning and Development Department) (1991) *Planning to Grow Old in Toronto*, Toronto: City of Toronto Planning and Development Department.

Pearce, D. (1989) *Blueprint for a Green Economy*, London: FoE.

Pedestrians Policy Group (1996) *Walking Forward: What Government and Local Councils Need To Do to Get People Walking*, London: Transport 2000 Trust.

Pell, B (1991) 'From the public to the private sector', in H. Thomas and P. Healey (eds) *Dilemmas of Planning Practice*, Aldershot: Avebury.

Pemberton, S. and Vigar, G. (1998) *Managing Change in a Fragmented Institutional Landscape: The Micro-politics of Transport Planning's 'New Realism' in Tyne and Wear*, Department of Town and Country Planning working paper, Newcastle: University of Newcastle upon Tyne.

Penoyre & Prasad in association with Elsie Owusu Architects and Audley English Associates (1993) *Accommodating Diversity: The Design of Housing for Minority Ethnic, religious and Cultural Groups*, London: Matrix Architects Feminist Co-operative.

Powergen (1996) *The CDM Regulations: A Design Risk Assessment Manual*, London: Blackwell.

Pickvance, C. (1977) 'Physical Planning and Market Forces in Urban Development' National *Westminster Bank Quarterly Review*, August; reprinted in Paris (1982) pp. 69–82.

Pred, A. (1977) *City-Systems in Advanced Economies*, London: Hutchinson.

Pred, A. (1983) 'Structuration and place: on the becoming of sense of place and structure of feeling', *Journal for the Theory of Spatial Behaviour*, vol. 13, pp. 45–68.

Pred, A. (1984) 'Place as historically contingent process: structuration and the time geography of becoming places', *Annals of Association of American Geographers*, vol. 74, pp. 279–97.

Price, C. and Tsouros, A. (eds) (1996) *Our Cities, Our Future*, Copenhagen: WHO.

RDC (Rural Development Commission) (1995) *Rural Challenge Bidding Guidance*, Salisbury: RDC.

Redfield, J. (1994) *The Celestine Prophecy*, London: Bantam.

Redwood, J. (1997) 'Time to let other options take root', *The Age* (Australia) 14 July, p. A13.

Rees, G. and Lambert, J. (1985) *Cities in Crisis: The Political Economy of Urban Development in Post-War Britain*, London: Arnold.

Revie, C. (1997a) 'Why greening the Maastricht Treaty is important for Scotland', *Scottish Affairs*, no. 20, Summer, pp. 95–111.

Revie, C. (1997b) *Achieving the Possible: A Sustainable Energy Strategy for Scotland*, Edinburgh: FoE Scotland.

Rhodes, R. (1997) *Understanding Governance*, Oxford: Oxford University Press.

Rhys Jones, S., Dainty, A., Neale, R. and Bagihole, B. (1996) *Building on Fair Footings: Improving Equal Opportunities in the Construction Industry for Women*. Proceedings of Construction Industry Board Conference.

RIBA (Royal Institute of British Architects) (1992) *Education Statistics* London: RIBA Glasgow: CIB.

Rickford, F. (1993) 'Wise heads', *Search*, Autumn, pp. 21–3.

Rodaway, P. (1994) *Sensuous Geographies*, London: Routledge.

Roberts, M. (1991) *Living in a Man-Made World: Gender Assumption in Modern Housing Design*, Routledge: London.

Root, A. (1996) 'Empowering local communities through transport', *Local Government Policy-Making*, vol. 22, no. 4, pp. 32–7.

RTPI (Royal Town Planning Institute) (1976) *Education Policy: Guidelines for Planning Schools*, London: RTPI.

RTPI (1985) *Access for Disabled People*, London: RTPI. Revised 1988 as Planning Advice Note (PAN) no. 3.

RTPI (1991a) *The Education of Planners: Policy Statement and General Guidance for Academic Institutions Offering Initial Professional Education in Planning*, London: RTPI.

RTPI (1991b) *The Building Regulations 1985 Part M: Access for Disabled People*, London: RTPI.

RTPI (1993) *Access Policies for Local Plans*, London: RTPI.

RTPI (1997) *Members' Survey: What Do We Need to Do To Ensure the Future of Planning and the Planning Profession?*, London: RTPI.

Rural Forum and Rural Research Branch (1997) *Community Involvement in Rural Development Initiatives: Good Practice in Rural Development*, no. 2, Edinburgh: Scottish Office Central Research Unit.

Russo, A., Mohanty, C.T. and Torres, L. (eds) (1991) *Third World Women and the Politics of Feminism*, Indiana: Indiana University Press.

Rutherford, A. (1998) 'Islay: a model for rural communities and beyond' in Scottish Office (1998) *No Small Change*, Edinburgh: Scottish Office.

Sachs, C. (1983) *Invisible Farmers: Women's Work in Agricultural Production*, Totowa, NJ: Rhinehart & Allenheld.

Sandercock, L. (1998) *Towards Cosmopolis*, Chichester: John Wiley.

Sandercock, L. and Forsyth, A. (1990) *Gender: A New Agenda for Planning Theory'*, Institute of Urban and Regional Development working paper 521, Berkeley, CA: University of California.

SAPN (Scottish Anti-Poverty Network) (1996) *The Scottish Declaration on Poverty: A Framework for Action Against Poverty by Central and Local Government, Health Boards, Enterprise Companies and the Private, Voluntary and Community Sectors*, Glasgow: SAPN.

Scott, A.J. and Roweis, S.T. (1977) 'Urban planning in theory and practice: a reappraisal' *Environment and Planning A*, vol. 9, pp. 1097–119.

Scottish Office (1988) *New Life for Urban Scotland*, Industry Department, Edinburgh: Scottish Office.

Scottish Office (1995a) *Programme for Partnership: Review of Urban Regeneration Policy*, Edinburgh: Scottish Office.

Scottish Office (1995b) *Rural Scotland: People, Prosperity and Partnership. The Government's Policies for the Rural Communities of Scotland*, Cmnd 3041, Edinburgh: Scottish Office.

Scottish Office (1995c) Progress in Partnership: A Consultation Paper on the Future of Urban Regeneration Policy in Scotland, Edinburgh: Scottish Office.

Scottish Office (1996) *The Scottish Environmental Protection Agency and Sustainable Development*, Edinburgh: Scottish Office.

Scottish Office (1997) *Coastal Planning*, National Planning Poliicy Guideline 13; Edinburgh: Scottish Office.

Scottish Office (1998a) *Working Together for a Healthier Scotland: A Consultation Document*, Cmnd 3584, Department of Health, Edinburgh: HMSO.

Scottish Office (1998b) *No Small Change: Sustainable Development in Scotland*, Edinburgh: Scottish Office.

SEF (Scottish Environmental Forum) (1997) *Poverty and Sustainable Development in Scotland: Meeting the Challenge of Implementing Chapter 3 of Agenda 21*, Edinburgh: SEF.

Select Committee on the Environment (1997) *Fourth Report: Shopping Centres*, government response, Cmnd 3729 London: DETR.

Selman, P. (1996) *Local Sustainability: Managing and Planning Ecologically Sound Places*, London: Paul Chapman.

Selman, P. H. and Barker, A. J. (1989) 'Rural land use policy at the local level: mechanisms for collaboration' *Land Use Policy*, vol. 6, no. 4, October, pp. 281–94.

SEPA (Scottish Environmental Protection Agency) (1996) *State of the Environment Report*, Stirling: SEPA.

Sewel, Lord (1997a) 'Central and local government in accord', speech to the conference 'Sustaining Change: Local Agenda 21 in Scotland', City of Edinburgh Council, City Chambers, Edinburgh, 21 November.

Sewel, Lord (1997b) 'The realities of change: sustainable development in Scotland' address to the ERM Scotland Environmental Forum, 18 September, Edinburgh: ERM and the Scottish Office.

Sheffield City Council (1988) *A City Centre for People: Sheffield Central Area Local Plan*, Sheffield: Sheffield City Council.

Shortall, S. (1992) 'Power analysis and farm wives: an empirical analysis of the power relationships affecting women on Irish farms', *Sociologia Ruralis*, vol. 32, no. 4, pp. 431–51.

Shotter, J. (1993) *Cultural Politics of Everyday Life*, Milton Keynes: Open University Press.

Simmie, J. (1974) *Citizens in Conflict: The Sociology of Town Planning*, London: Hutchinson.

Simmons, D.G. (1994) 'Community participation in tourism planning', *Tourism Management*, vol. 15, no. 2, pp. 98–108

Skjerve, R. (ed.) (1993) *Manual for Alternative Municipal Planning*, Oslo: Ministry of the Environment.

SNH (Scottish Natural Heritage; Historic Scotland) (1996) *Environmentally Sensitive Farming in the Argyll Islands: A Practical Guide for Farmers, Crofters and Grazing Communities*, Perth: SNH.

SOBA (Society of Black Architects) (1997) 'Mentoring: to tame or to free?' unpublished symposium notes circulated at meeting of Society of Black Architects, 27 November, Prince of Wales's Institute of Architecture, London.

Social Europe (1996) 'Europe in the twilight zone', *Social Europe*, November, pp. 21–2.

SODD (Scottish Office Development Department) (1997) *Planning and Environmental Protection*, Planning Advice Note 51, Edinburgh: HMSO.

Sowerby, H. (1994) 'Local Pensioners' Forum', *North East Pensioners' Voice*, Newcastle: North East Pensioners' Association, June/July, 8–9.

Spitzner, M. (1998) 'Distanz zu Leben, Arbeit und Gemeinschaft?', paper presented at the *Eurofem conference*, 'Local and Regional Sustainable Human Development from a Gender Perspective', 10 June, Hämeenlinna, Finland.

Spitzner, M. and Zauke, G. (1996) 'Evaluation of the involvement of women in transport science' summary paper presented at 'Politiche del Tempo in una Prospettiva Europea: Conferenza Internazionale'. Eurofem Conference, Aosta, Turin, September 1996 (contact Wuppertal Urban Institute).

Spitzner, Meike (1998) 'Distanz zu Leben, Arbeit und Gemeinschaft?' paper presented at the *Eurofem, Gender and Human Settlements Conference* on Local and Regional Sustainable Human Development from a Gender Perspective, 10 June, Hämeenlina, Finland.

Stark, A. (1997) 'Combating the backlash: how Swedish women won the war' in A. Oakley and J. Mitchell (eds) (1997) *Who's Afraid of Feminism? Seeing Through the Backlash*, London: Hamish Hamilton, pp. 224–44.

Stoker, G. and Young, S. (1993) *Cities in the 1990s*, Harlow: Longman.

Stone, M. (1990) *Rural Childcare*, Rural Development Commission Research Report no. 9, London: Rural Development Commission.

Swain, J., Finkelstein, V., French, S. and Oliver, M. (eds) (1993) *Disabling Barriers: Enabling Environments*, Milton Keynes: Open University.

Swindon Access Action Group (1997) *Designing a Barrier-Free Environment: A Planning Equality Guidance Note*, Swindon: Swindon Access Action Group.

Swindon Access Action Group (1998) *A Proposal for A Strategy for Equality*, Swindon: Swindon Access Action Group.

Swindon Borough Council (1996) *Thamesdown Local Plan, Public Local Inquiry: Report of the Inspector and Report of the Summary of Recommendations*, Swindon: Swindon Borough Council.

Swindon Borough Council (1997/8) *Minutes of Railway Museum Access Advisory Group*, Swindon: Swindon Borough Council.

Swindon Borough Council (1998) *Local Plan Deposit Draft: List of Proposed Modifications*, Swindon: Swindon Borough Council.

Szinovacs, M. (1982) (ed.) *Women's Retirement: Policy Implications of Recent Research*, London: Sage.

Taylor, N. (1998) *Urban Planning Theory Since 1945*, London: Paul Chapman.

Tewdwr-Jones, M. (1995) *British Planning Policy in Transition*, London: Taylor and Francis, UCL Press.

Thamesdown Borough Council (1995a) *Access Design for Life: Accessible Housing in Practice*, Swindon: Thamesdown Borough Council, Community Development Department.

Thamesdown Borough Council, (1995b) *Draft Building Regulations for Access into New Dwellings*, Swindon, Thamesdown Borough Council.

Thomas, H. (1980) 'The education of British town planners, 1965–1975', *Planning and Administration*, vol. 7, no. 2, pp. 67–78.

Thomas, H. (1984) 'British town planners under the Thatcher government', *Plan Canada*, vol. 24, no. 2, pp. 63–74.

Thomas, H. (1992) 'Volunteers, involvement in planning aid' *Town Planning Review*, vol. 63, no. 1, pp. 47–62.

Tickell, A. and Peck, J. (1995) 'Social regulation after Fordism: regulation theory, neo-liberalism and the global-local nexus', *Economy and Society*, vol. 24, no. 3, pp. 357–86.

Tickell, A. and Peck, J. (1997) 'The return of the Manchester men: men's words and men's deeds in the remaking of the local state', *Transactions of the Institute of British Geographers*, vol. 21, no. 4, pp. 595–616.

Trioli, V. (1997) 'In Gippsland's forests it is a daily battle of wit and tactics', *The Age* (Australia), 21 July, pp. A1, A6.

Turner, G. (1992) 'Public/private sector partnership: panacea or passing phase?', *Insights*, March, pp. A85–A91.

Tuxworth, B. and Thomas, E. (1996) *Local Agenda 21 Survey 1996: Results May 1996*, London: LGMB (Local Government Management Board).

UK Round Table on Sustainable Development (1996) *Defining A Sustainable Transport Sector* London: HMSO.

UK Representation (1997) *European Social Policy*, London: UK Representation of the European Commission.

UK Steering Group (1994) *Local Agenda 21: Principles and Process*, London: UK Steering Group.

UN (United Nations) (1997) Kyoto Protocol to the UN Framework Convention on Climatic Change, December 1997, document reference FCCC/CP/1997/L.7/Add. 1. New York: United Nations.

UNCED (United Nations Conference on Environment and Development) (1992) *Earth Summit*, Press summary of Agenda 21, Rio de Janeiro: UNCED.

Unwin, T. (1998) 'Agricultural change and rural stress in the new democracies', in D. Pinder (ed.) *The New Europe*, London: Wiley.

van den Berg, L. (ed.) (1994) *Urban Tourism*, Rotterdam: Erasmus University.

Vigar, G. (1996) 'A new paradigm for transport planning?', paper presented at ACSP/AESOP conference, Toronto, July.

Vigar, G., Healey, P., Hull, A.D. and Davoudi, S. (forthcoming) *Planning and Place-making: Spatial Strategy and the English Planning System*, London: Macmillan.

WAC (Women Architects' Committee) (1993) *Women Architects*, London: RIBA.

Walton, W., Ross, A. and Rowan-Robinson, J. (1995) 'The precautionary principle and the UK planning system', *Journal of Planning and Environmental Law*, pp. 986–93.

Ward, S. V. (1994) *Planning and Urban Change*, London: Paul Chapman.

WCC (Women's Communication Centre) (1996) *Values and Visions: The 'What Women Want' Social Survey*, London: WCC.

WCED (World Commission on Environment and Development) (1987) *Our Common Future* (The Brundtland Report), Oxford: Oxford University Press.

WDS (Women's Design Service) (1997) *The Good Practice Manual on Tenant Participation*, written by M. Kelly and C. Clarke, London: Women's Design Service.

Weber, M. (1964) *The Theory of Social and Economic Organisation (Wirtschaft und Gesellschaft)* ed. T. Parsons, New York: Free Press.

Wellbank, M. (1994) 'Sustainable development', proceedings of Town and Country Planning Summer School 1994, *Planning*, RTPI conference supplement, pp. 13–17.

WGSG (Women and Geography Study Group) (1997) *Feminist Geographies: Explorations in Diversity and Difference*, Harlow: Longmans, London: Women and Geography Study Group of Royal Geographical Society and Institute of British Geographers.

Whatmore, S. (1991) *Farming Women: Gender, Work and Family Enterprise*, London: Macmillan.

Whitelegg, J. (1997) *Critical Mass*, London: Pluto Press.

WHO (1990) *Making Partners: Intersectoral Action for Health*, Copenhagen: WHO.

WHO (1994) 'Constitution of the World Health Organisation', in *WHO Basic Documents*, 40th edition, Geneva: WHO.

WHO (1997) *City Planning for Health and Sustainable Development*, Copenhagen: WHO.

Wilcox, S. (1998) *Housing Finance Review 1997*, York: Joseph Rowntree Foundation.

Wightman, A. (1996) Scotland's Mountains: An Agenda for Sustainable Development, Perth: Scottish Wildlife and Countryside Link.

Wilkinson R. and Marmot M. (eds) (1998) *The Solid Facts: Social Determinants of Health*, Copenhagen: WHO.

Williams, R. (1961) *The Long Revolution*, Harmondsworth: Penguin.

Williams, R. H. (1975) 'The idea of social planning', *Planning Outlook*, pp. 11–18.

Williams, R. H. (1996) *European Union Spatial Policy and Planning*, London: Paul Chapman.

Williams, R. H. (1997) 'A happy end? Prospects for the ESDP under the British presidency' *Built Environment* vol. 23, no. 4, pp. 315–8.

Williams, R. H. (1998) 'The next steps in the EU spatial agenda', in C. Bengs and K. Böhme (eds) *The Progress of European Spatial Planning*, Stockholm: NORDREGIO.

Winkler, T. and Mitchell, B. (1997) 'Greens vow to fight regional forest', *The Age* 4 February, p. A6.

Woo, M. (1981) 'Letter to Ma', in C. Moraga and G. Anzaldua (eds) *This Bridge Called my Back*, London: Kitchen Table; Women of Colour Press.

Wood, J., Gilroy, R., Healey, P. and Speak, S. (1995) *Changing the Way We Do Things Here*, Newcastle: Centre for Research on European Urban Environments, University of Newcastle.

Wood, P. (1996) 'Regeneration and the arts in Kirklees' in F. Matarasso and S. Halls (eds) *The Art of Regeneration*, Nottingham and Bournes Green: City of Nottingham and Comedia.

Wright, K. (1995) 'Help save the dunes' letter to the editor, *Wanneroo Times* (Australia), 7 November.

Young, M. and Willmott, P. (1957) *Family and Kinship in East London*, Harmondsworth: Penguin (revised edition).

Scottish Office publications of relevance to social town planning

1988   *New Life for Urban Scotland*, Industry Department for Scotland.

1994   *The Planning System*, National Planning Policy Guideline 1.

1995   *Programme for Partnership: Review of Urban Regeneration Policy.*

1995   *Progress in Partnership: A Consultation Paper on the Future of Urban Regeneration Policy in Scotland.*

1995   *Rural Scotland: People, Prosperity and Partnership. The Government's Policies for the Rural Communities of Scotland*, Cmnd 3041.

1996   *Transport and Planning Consultative Draft*, National Planning Policy Guideline.

1996   *The Scottish Environment Protection Agency and Sustainable Development.*

1996   *Programme for Partnership: Guidance on applying for Urban Programme Funding.*

1997   *Coastal Planning*, National Planning Policy Guideline 13.

1997   *Community Involvement in Rural Development Initiatives : Good Practice in Rural Development*, No. 2, Central Research Unit, Rural Forum and Rural Research Branch.

1997   *Planning in Small Towns.*

1997   *Keeping Scotland Moving*, Transport Green Paper.

1997   *Common Sense, Common Purpose: Sustainable Development in Scotland.*

1997   *Towards a Development Strategy for Rural Scotland: Discussion Paper.*

1998   *Social Exclusion: Consultation Paper.*

1998   *Working Together for a Healthier Scotland: A Consultation Document*, Cmnd 3584, Department of Health.

1998d *No Small Change: Sustainable Development in Scotland.*

WWW sites of relevance to sustainable development issues

The number of sites is expanding all the time; those listed below offer an introduction for surfers. All sites have links to associated sites.

Agenda 21
http: //www.agenda21.se/

CADISPA
http: //www.strath.ac.uk/Departments/CommunEdu/

Earth Council: an international NGO to promote implementation of the Earth Summit (1992)
http://www.ecouncil.ac.cr/

Environmental Resources: link to research sites and publications from UK, Europe and rest of world
http://www.geog.le.ac.uk/cti/resman.html

European Union: index site for EU, including its institutions and policy documents
http://www.europa.eu.int/

Friends of the Earth Scotland
http: //www.foe-scotland.org.uk

Labour manifesto
http: //www.labour.org.uk/core.html

RESET: renewable energy strategies for European towns
http://resetters.org/resetnet/index.htm

RSPB
http: //www.rspb.org.uk

RTPI
http: //www.rtpi.co.uk

SEPA
http: //www.sepa.org.uk

SURBAN: Sustainable Urban Development in Europe; good practice examples
http://www.eaue.de/winuwd/

Sustainable Development for Local Authorities: European Municipal Associations
http://www.eea.dk/Projects/EnvMaST/susdevla/ 2euminas.htm

UK government: index site for UK government departments, including DETR and the Scottish Office
http://www.open.gov.uk/

WWF
http: //www.panda.org/

Name index

Subject index